Beach-Spawning Fishes

Reproduction in an Endangered Ecosystem

Beach-Spawning Fishes

Reproduction in an Endangered Ecosystem

Karen L. M. Martin

Pepperdine University
Malibu, California, USA

CRC Press
Taylor & Francis Group
Boca Raton London New York

CRC Press is an imprint of the
Taylor & Francis Group, an **informa** business

CRC Press
Taylor & Francis Group
6000 Broken Sound Parkway NW, Suite 300
Boca Raton, FL 33487-2742

First issued in paperback 2020

ISBN-13: 978-1-4822-0797-2 (hbk)
ISBN-13: 978-0-367-65905-9 (pbk)

This book contains information obtained from authentic and highly regarded sources. Reasonable efforts have been made to publish reliable data and information, but the author and publisher cannot assume responsibility for the validity of all materials or the consequences of their use. The authors and publishers have attempted to trace the copyright holders of all material reproduced in this publication and apologize to copyright holders if permission to publish in this form has not been obtained. If any copyright material has not been acknowledged please write and let us know so we may rectify in any future reprint.

Visit the Taylor & Francis Web site at
http://www.taylorandfrancis.com

and the CRC Press Web site at
http://www.crcpress.com

This book is dedicated to Doug Martin,
always my favorite Pisces and the best catch of any day.

Contents

Preface

The natural world has certain rules. The trouble is, life has a way of bending those rules. Fishes slither and crawl about out of water without legs. Tiny anamniotic eggs incubate on land. Fishes without lungs emerge into air. Welcome to the intriguing world of beach-spawning fishes.

My adventures with beach-spawning fishes started on a California beach late at night, surrounded by friends and family. The froth of a receding wave suddenly came alive with dancing silverside fishes. The California grunion were running, forming a living mirror of the starlit sky above. Having studied air breathing and emergence of many species of fishes from the rocky intertidal zone, I began looking for similarities to those tide pool fishes. I intended to spend a couple years pursuing questions about California grunion and then move on. Nearly 20 years later I continue to be fascinated by this species, and I still have more questions than answers, because apparently that is the way science progresses.

The genesis of this book was the realization that, spectacular as the California grunion are, they are not the only fishes that lay eggs on beaches. My first impression that this was an extremely rare, possibly unique behavior has been revised in recognition of the many diverse examples of beach-spawning fishes. Some tide pool fishes with seemingly quiet lives engage in exotic sexual behavior in and out of water. Some fishes that live in deeper water migrate many miles just to mate and set up a nursery on shore. With a comparative approach, it becomes clear that beach-spawning fishes take many different shapes and may be found in myriad families, on any type of beach, in many parts of the world.

These fishes tend to be small and inconspicuous. Only a few are economically important, but all contribute to the marine food web and connect the nutrient cycles of the land to the sea. Like many small, short-lived animals, their populations may dramatically expand or contract in different years. These oscillations may be natural, but they may be altered by pressures from human activities, either directly from fishing or indirectly from habitat loss or degradation.

Beaches are some of the most beautiful places on Earth. The economic importance of beaches for recreation, residences, and businesses underscores the urgency for protection of the natural ecological functions that beaches can retain even in the most urbanized settings. A habitat that transitions between ecosystems is vulnerable to changes from either side. Coastal development on a global scale squeezes critical habitat already facing sea level rise. Shoreline armoring changes natural contours and erosion and deposition of sand or gravel. Dams reduce the replenishment of sediments on shore. Vehicles driven on beaches may crush cryptic nests or incubating embryos. Beach-spawning fishes are vulnerable to all of these individual and cumulative impacts.

Fortunately, these unassuming animals, with their unexpected lifestyles, inspire people to protect their critical habitats. Over the past few decades, many groups have organized to monitor and defend species of beach-spawning fishes on local and regional levels. New policies and heightened awareness may increase protection of critical habitat and vulnerable spawning fishes, but surely more can be done. The involvement of the public is vital for the success of these initiatives; their careful observations and boundless enthusiasm are irreplaceable.

The purpose of this book is to enlighten, energize, and empower people to increase our scientific knowledge and to preserve the ecological functions of our beaches as coastal natural habitats. This book is for these exuberant beach-spawning fishes, with passion that even the vastness of the ocean cannot contain.

Acknowledgments

I thank Charles Crumly for his continuing interest in life at the water's edge, and the good folks at Taylor & Francis for this opportunity to explore it further.

I am grateful for cheerful and extensive conversations with many scientists about their research over the long course of development of this book. Michael Horn, Karen Warkentin, Jeff Graham, Don Thomson, Andy Olson, Jr., Doug Middaugh, Daniel Penttila, Kazunori Yamahira, Richard Strathmann, Christopher Petersen, Roger Seymour, David Bradford, Patricia Wright, Gianluca Polgar, Kenneth and Gabrielle Addman, Bob Lea, Dale Roberts, Lynne Parenti, Milton Love, Atsushi Ishimatsu, Leonard DiMichele, Malcolm Taylor, Kenneth Frank, Camm Swift, Andrew Kinziger, Michael Schaadt, Suzanne Lawrenz-Miller, John Lighton, Brian Nakashima, Anna Olafsdottir, Kathryn Hieb, Andy Jahn, Christopher Bridges, Claus Dieter Zander, Jurgen Nieder, Jason Podrabsky, Ed DeMartini, Jeff Marliave, Joe Cech, Danielle Zacherl, Ron Burton, Rodney Honeycutt, Lee Kats, Tom Vandergon, Donna Nofziger Plank, Kathryn Dickson, Don Swiderski, Giacomo Bernardi, Don Buth, Larry Allen, Nancy Aguilar, Julianne Steers, Andres Carrillo, Ben Higgins, Hannes Baumann, Elizabeth Brown, Helena Aryafar, Helen Hess, Sulayman Mourabit, Ben Crabtree, Richard Rosenblatt, Jules Crane, and Boyd and Mary Walker generously shared their expertise in coastal fishes and eggs out of water.

The following people helped me to appreciate beaches as ecosystems and to understand the threats facing them. I am grateful to Jenny Dugan, Dave Hubbard, Bill Patzert, Dennis Simmons, Dennis Lees, Dennis Reed, Dave Pryor, Peter Brand, Rosi Dagit, Suzanne Goode, Anton McLachlan, Bill Beebe, Nick Steers, Nancy Aguilar, Mark Gold, Shelley Luce, Rich Ambrose, Loni Adams, Jerry Schubel, Marilyn Fluharty, Debbie Aseltine, Ed Roberts, Sarah Sikkich, Steve Murray, Jonna Engel, John Dixon, Leslie Ewing, Bob Grove, Bob Hoffman, Bryant Chesney, Chad Nelsen, Rick Wilson, Nancy Hastings, Phil King, Andrew Staines, Stacey Vigallon, Tom Ryan, Dana Murray, Aaron McLendon, Pat Veesart, Adam Obaza, Guangyu Wang, Mike Rouse, Ken Lohmann, Kenneth Weiss, Sean Anderson, Travis Longcore, Harry Helling, and Peter Douglas.

Many intelligent, enthusiastic student collaborators have taught me so much, in particular Cassadie Moravek, Rachel Darken, Meredith Fisher, Harris Lakisic, Josh Drais, Christopher Van Winkle, Jessica Hernandez, Erica Robbins, Ariel Carter, Avery Powell, Karen Bailey, Rachael Pommerening, Tara Speer Blank, Rebecca Ashley, Stacy DiRocco, Kate Mulder, Ryun Harper, Raquel Rausch, Colin Byrne, Eian Carter, Jeremiah Bautista, Jennifer Griem Raim, Jennifer Flannery Harr, Kim Carpenter Misamore, Beth Smyder Tramontina, Michael Lawson, Michael McClure, Kjirsten Carlson, Alice Walker, Phil Johnson, Juli Matsumoto, Anastasia Fry, Christopher Hakim, Sarah Clark, Kristie Haney, Devin Sanders, Hilary Hays Engebretson, Chantil Ayres Ruud, Aimee Luck, Eddie Mah, Ryan Bloom, Mia Holmes, Vincent Quach, and Emily Pierce.

Words cannot express my gratitude for the efforts and enthusiasm of the thousands of Grunion Greeters in California and Baja California, and for the constant guidance and support of Melissa Studer, Dennis Simmons, and Bob Hoffman. I am grateful for the cooperation of many conservation organizations, including Cabrillo Marine Aquarium, Birch Aquarium at Scripps, Aquarium of the Pacific in Long Beach, Ocean Institute in Dana Point, Surfrider Foundation, Heal the Bay, Santa Barbara Channelkeepers, the Audubon Society, California State Parks, Monterey Bay Aquarium Research Institute, and so many others.

Research funding was provided by many sources, including the National Science Foundation, the National Marine Fisheries Service, the National Oceanic and Atmospheric Administration, California Sea Grant, University of Southern California Sea Grant, National Fish and Wildlife Foundation, California Coastal Commission, National Geographic Society Committee on Research and Exploration, the Tooma Foundation, the Coastal America Foundation, and Pepperdine University. The book was written during a much appreciated sabbatical leave.

For their sensitive artwork, I thank Andrea Lim and Gregory Martin. Bill Hootkins provided bountiful hospitality at his Malibu beach house along with his distinctive photographs. I am grateful to Tim Bourque, Ken Frank, Jennifer Flannery Harr, Chris Lindeman, Anna Olafsdottir, Doug Martin, Peter Heistand, Joe McLain, and Hikari City in Japan for permission to use photographs; and to Doug Middaugh, Norio Shimizu, Don Swiderski, Mike Schaadt, and Nils Stenseth for figures.

All guidance and suggestions were provided with kindness and tact. Any errors or omissions are completely my own responsibility. I thank Jean Matthews, Chris Petersen, Andy Olson, Jr., Dan Penttila, Rodney Honeycutt, Doug Martin, and Alex Martin for perceptive comments on portions of the manuscript. I can never repay my infinitely patient graduate advisors, Vic Hutchison at Oklahoma University and Ken Nagy at the University of California, Los Angeles, for imparting a strong foundation in physiological ecology and a fascination with unexpected solutions to life's challenges.

Introduction

Spawning on beaches by fishes is an unexpected form of reproduction that is seen in a surprising number of teleost fish species. This charismatic animal behavior provides a window into the ecological conditions and organismal physiology necessary for aquatic fishes and their eggs to adapt to terrestrial conditions. The importance of this critical habitat for the survival of these species highlights some of the threats and challenges for conservation and management in coastal areas.

New challenges await the adults and embryos during and after intertidal spawning. They emerge from water during low tides and may be exposed to novel avian and terrestrial predators that never hunt in deeper waters. The physiology, behavior, anatomy, and ecology of spawning fishes and their care of nests and incubating embryos are explored in depth to better understand the adaptive advantages and disadvantages.

Renewed scientific interest in many beach-spawning fishes over the past few decades has resulted in a rich literature that offers new insights into the biology and ecology of beach-spawning fishes. The many beach-spawning species in diverse lineages of teleost fishes exhibit a wealth of adaption to the ever-changing habitats at the ocean's edge. Marine fishes show a variety of ecological triggers for air-breathing and amphibious behavior, leaving the water for reproduction, parental care, feeding, or escaping harsh aquatic conditions. Cues for emergence and air breathing are different for amphibious marine fishes than for air-breathing fishes from freshwater.

This book provides the first detailed scholarly exploration of an iconic type of behavior in fishes. Beach-spawning fishes have fascinated people for centuries. Spawning runs of lively fishes, emerging as if by magic out of ocean waves, have inspired art, music, poetry, television, and film, as well as many romantic rendezvous for the human species, and some unique opportunistic fisheries. This is the first comprehensive survey of an extensive literature on natural history, behavior, ecophysiology, and developmental biology. New knowledge about these beach-spawning fishes may spark greater interest in identifying and protecting them globally.

Beach-spawning species include a surprisingly large number of fishes that reproduce on land, including species that spawn in tide pools, on sandy beaches, in estuaries, and in other coastal habitats throughout the world. Some examples are important forage fishes such as the Canadian capelin, exotic creatures such as the toxic kusafugu puffer of Japan, and the spectacular midnight surfers of the Pacific coast, the California grunion. Beach-spawning fishes occur in many different families and on coasts across the entire globe. See Chapter 1 for an evolutionary and biogeographic overview.

The essential spawning habitat of each beach-spawning species is a narrow ribbon of coastline that contains a particular substrate, beaches made of sand, gravel, rock, or mud. The appropriate habitats are patchily distributed along coastlines. In most cases, tidal excursions allow exposure of developing embryos to air for some or all of their incubation period. The physical conditions of beach ecosystems are described in Chapter 2.

The spawning runs of teleost fish species may be regular and predictable or tightly synchronized with high semilunar tides. Runs may take place under cover of darkness, long after humans have left the shore. Some spawning events occur in remote locations or take place so infrequently that they are unlikely to be seen by human observers. In addition to fishes that reside intertidally, many subtidal teleosts and a few diadromous fishes migrate to beaches for reproduction, suggesting that beach-front properties are desirable nurseries even for nonresident species. Parents may attach clutches of eggs to rocks or vegetation and then leave after spawning, or one or both parents may stay and guard the nest during development. This parental behavior may leave the fish out of water as the tide ebbs. See Chapters 3 and 4 for specific details. In addition, a few species of fishes that live in freshwater spawn in places or at times that result in their eggs being emerged from water during incubation.

Although many of these fishes emerge into air either while spawning or during parental care, most exhibit no dramatic morphological adaptations for terrestrial emergence or air breathing. Some exchange oxygen and carbon dioxide in both water and air with ease. Chapter 5 explores the connections between beach spawning and air breathing in teleost fishes and their embryos.

The spawning aggregation of like-minded fishes that arrive on the appropriate substrate at the right tidal height is impressive to behold. They literally risk their lives to rush to the water's edge and lay their eggs high on the shore, exposing themselves and their offspring to a gauntlet of marine, terrestrial, and avian predators. With beach spawning as a life strategy, embryos and vulnerable adults must cope with natural threats such as predators and exposure to harsh conditions from both marine and terrestrial influences in this edge habitat. See Chapter 6 for more on unsafe sex and predation during beach spawning and incubation on shore.

The intertidal zone of the beach is a transition between seawater and air. This edge habitat makes a grand nursery for fish eggs, with easy availability of moisture and oxygen, temperatures that may be warmer than the surrounding ocean, and few marine predators. Embryos of fishes developing in the variably submerged and emerged beach habitat are produced in large, yolky, demersal eggs that resist desiccation. See Chapter 7 for details on tidally influenced incubation and early life in these fishes.

Since hatching on land would be harmful to fishes, some species have environmentally cued hatching and the ability to extend incubation until favorable conditions arise. Embryos from the same spawning date may hatch at different times and develop through larval stages on different dates even though they are the same chronological age. Tidal emergence delays hatching in some species, and submergence cues hatching in several species. See Chapter 8 for more on the types and triggers of hatching in embryos of fishes that incubate on beaches.

Many beaches that are critical habitat for fish reproduction are intensively managed for human recreation, visited by millions of people every year. In some areas, runs of beach-spawning fishes have become popular natural phenomena, attracting wildlife watchers by the thousands to observe and enjoy the remarkable behaviors. New studies on the dynamic, shifting substrates of the beach habitat in recent years allow integration of different beach zones and greater understanding of biological interactions among beach denizens. The importance of beaches as nursery areas for many different species of birds, reptiles, mammals, and arthropods is now appreciated to a greater extent than in the past.

Anthropogenic impacts, both intentional and unintentional, on beach-spawning fishes include recreational fishing, coastal construction, shoreline armoring, beach cleaning practices, and changes in water quality. Chapter 9 describes some of the emerging threats to beach ecosystems. Human activities along the coast and changing climates are likely to lead to increased vulnerability of early life cycle stages of beach-spawning fishes. These influences could affect geographic distribution and population structure of beach-spawning species in ways that differ from the effects on more typical, completely aquatic fish species. It is well understood that many mammalian and reptilian species that reproduce on beaches are endangered, struggling to survive in a changing habitat heavily influenced by human activities.

Raising public awareness of beaches as important marine ecosystems increases ecological sensitivity for beach management, enhancing the habitat for additional diverse species. Efforts by citizens, aquariums, and environmental groups to protect an endemic beach-spawning fish in California have changed the way that beaches and coastal areas are managed. The beauty, intriguing biology, and importance of these charismatic fishes at the interface between marine and terrestrial ecosystems have inspired efforts to protect their coastal nurseries. Chapter 10 discusses some of the grassroots conservation efforts that have sprouted to address concerns about vulnerable beach-spawning fish species and their spawning runs. New policies to protect the spawning grounds and regulate fisheries are helping these species to continue into the future.

Beach-spawning fishes provide spectacular examples of extreme adaptations during the most vulnerable life cycle stages, in exotic locations on most continents of the world. The distinction between marine and terrestrial life is more permeable than was previously thought. These habitats are bridged by many fishes from the very beginning of their lives. In an era of climate change, we humans have much to ponder about these exuberant creatures in this endangered habitat.

About the Author

Karen L. M. Martin, PhD, is professor of biology and Frank R. Seaver Chair in Natural Science at Pepperdine University in Malibu, California. She is a fellow of the American Institute of Fishery Research Biologists, and active in the American Fisheries Society, the American Society of Ichthyologists and Herpetologists, the Society for Integrative and Comparative Biology, and other scientific organizations. She has published extensively on studies of animals at the interface between water and land and is associate editor of the journal *Copeia*. Dr. Martin's previous books are *Amniote Origins Completing the Transition to Land*, coedited with Stuart Sumida; and *Intertidal Fishes: Life in Two Worlds*, coedited with Michael Horn and Michael Chotkowski.

After earning undergraduate and graduate degrees at Oklahoma University, she earned a doctorate in biology at Universtiy of California, Los Angeles, and completed a postdoctoral Friday Harbor Fellowship at the University of Washington. Dr. Martin received the Environmental Partnership Award from the American Shore and Beach Preservation Association, and the Conservation Achievement Award from the American Fisheries Society. Her award-winning short documentary, *Surf, Sand, and Silversides: The California Grunion*, has been screened at numerous regional and international film festivals, conferences, universities, and aquariums. She cofounded the Beach Ecology Coalition and the Grunion Greeters in order to work with multiple stakeholders on beaches to balance human recreation with wildlife conservation. Dr. Martin is an advocate for involving the public in scientific research.

1 A Leap of Faith: The Evolution of Beach Spawning in Fishes

Careful examination of the lives of real animals in the real world reveals the true complexities and dangers involved in the seemingly simple tasks of living and reproducing. These are completed only by the successful few. For most fishes, and indeed most animals, those that survive to adulthood are a tiny fraction of the offspring produced.

Fishes that spawn at the water's edge leave the eggs with their developing embryos high and dry on shore for some or all of their incubation. Fish species from many different lineages have been observed to reproduce on beaches. Even so, this behavior remains relatively rare and unusual for these fundamentally aquatic organisms. Of the nearly 500 families of fishes that exist in the waters of the Earth (Nelson, 2006), only 21 families are known to include species with beach spawning behavior (Table 1.1). Rare as it is, evidence to date indicates multiple independent origins of beach spawning among many different clades of teleost fishes, convergent behavior rather than shared evolutionary history.

This chapter considers what types of fishes are currently spawning on beaches, and what antecedent behaviors and reproductive strategies are most likely to have preceded the initiation of this behavior. This topic deserves scrutiny as it has been relatively unexplored in the past.

1.1 OVERVIEW OF BIOGEOGRAPHY AND ECOLOGY OF BEACH-SPAWNING FISHES

Biogeography reveals that beach-spawning fishes occur worldwide, in tropical, temperate, and circumpolar environments. Fishes spawn on beaches as far north as Scandinavia, Alaska, and the Sea of Okhotsk, and as far south as New Zealand and Tierra del Fuego. Some species spawn on tropical mudflats in Australia, Africa, and India, others on cool temperate gravel beaches in Canada and the Pacific Northwest. There are beach-spawning fishes on rocky coasts of Oregon and Japan, the Mediterranean Sea and Hawaii. Beach spawning fishes can be found on the sandy beaches of southern California, the seagrass meadows of Virginia, and the kelp beds in Prince William Sound. Beach spawning fishes can be found in China, Africa, the Mediterranean Sea, Egypt, Indonesia, and parts of South America. Species of beach spawning fishes have been found on the shores of all continents except Antarctica.

TABLE 1.1

At Least 21 Teleost Fish Families Are Known to Include Beach-Spawning Species

Superorder	Order	Family	Total Species	Beach-Spawning Species
CLUPEOMORPHA	CLUPEIFORMES	Clupeidae	200	1
PROTACANTHOPTERYGII	OSMERIFORMES	Osmeridae	15	4
		Galaxiidae	50	1
	SALMONIFORMES	Salmonidae	222	1 population of 1 species
PARACANTHOPTERYGII	BATRACHOIDIFORMES	Batrachoididae	78	1
ACANTHOPTERYGII	ATHERINIFORMES	Atherinopsidae	104	4
	CYPRINODONTIFORMES	Fundulidae	46	7
		Anablepidae	16	1
	GASTEROSTEIFORMES	Gasterosteidae	16	1 population of 1 species
	SCORPAENIFORMES	Cottidae	about 300	10+ intertidal species, 2 subtidal
		Liparidae	361	1 (there are probably more)

PERCIFORMES	Ammodytidae	17	1 (may be 2 species)
	Zoarcidae	220	1
	Pholidae	15	at least 4
	Stichaeidae	70	5
	Blenniidae	400	many intertidal and supralittoral
	Tripterygiidae	150	1 studied but others likely
	Gobiidae	2,000+	15 or more
GOBEISOCIFORMES	Gobiesocidae	161	many intertidal
TETRAODONTIFORMES	Tetraodontidae	120	1
PLEURONECTIFORMES	Pleuronectidae	103	1 population of 1 species

Sources: Classification is from Nelson (2006) and total species numbers for each family are taken from Froese & Pauly (2013).

Note: See Chapters 3 and 4 for more about the spawning behavior of each species.

The diversity of tropical, temperate, and polar habitats is matched by the diversity of spawning substrates favored by different species of beach-spawning fishes. Different species prefer vegetation such as seagrass or macroalgae, hard surfaces such as shells or boulders, or sediments from mud to coarse, well-drained sand to gravel. Spawning may occur in dynamic surge channels or in quiet estuaries, in a hidden chamber protected by a single male, or in a wave of an explosive mating assemblage with dozens of suitors.

Convergence of behavior generally implies a response to similar environmental factors, but this does not provide much insight for fishes that spawn on beaches because of the diversity of habitats in which beach-spawning occurs, and of the adult habitats. Some fishes that spawn on beaches live in estuaries or the rocky intertidal zone; others live offshore in subtidal waters. A few beach-spawning fish species are anadromous, and a very few species spawn at the edge of freshwater habitats. About the only similarities present in all of these habitats is the presence of a coastline, and the incorporation of tidal cycles to periodically expose the developing eggs and embryos to air. As discussed in Chapter 7, the survival of the embryos is related to their tidal height during incubation, and emersion into air increases oxygen availability as well as temperature variability. Desiccation and heat stress are the two major enemies of embryos emerged into air, with predation also a factor (see Chapter 6).

1.2   FISH FAMILIES WITH MANY INTERTIDAL SPECIES ALSO HAVE MANY SPECIES THAT SPAWN ON BEACHES

With few exceptions these species spawn in the same habitat as they live, so intertidal species spawn in the intertidal zone, and most subtidal species typically spawn in deeper water. However, some species of subtidal fishes spawn during high tides and place their eggs in the intertidal zone, where they will occasionally be emerged during low tides. The benefits the embryos of these species gain must outweigh any difficulties they encounter along the way at this earliest life stage.

The very speciose families of the Cottidae, Blenniidae, and Gobiidae contain numerous subtidal and intertidal species. These families have much to tell us about the evolution of spawning behavior and parental care (Almada & Santos, 1995; DeMartini, 1999; Petersen & Hess, 2011). In general these families all have demersal eggs that adhere to a substrate or vegetation of some sort, or to one another. This allows the spawning adults to perform at least rudimentary parental care in the form of selecting an appropriate substrate in either habitat, whether or not the clutch will be guarded after spawning.

In the evolution of beach-spawning behavior, adaptations to intertidal life for adult fishes probably evolved along with spawning in the intertidal zone (Horn et al., 1999). Comparisons of air-breathing ability across the family Cottidae show that emergence requires adaptations that are present in intertidal species but not in their closely related subtidal relatives (Martin, 1996; Martin & Bridges, 1999; Richards, 2011). A subtidal cottid, the sharpnose sculpin *Clinocottus acuticeps*, spawns in the rocky intertidal zone beneath rockweed (Marliave, 1981). A parent guards the nest during immersion, but not during low tides when the site is emerged into air.

Any embryos in eggs that develop in the intertidal zone will be emerged from water for at least some of their incubation period, usually in humid, protected locations such as under vegetation or shaded by boulders. A careful examination across the beach-spawning cottids might also reveal changes in the chorion structure or permeability, or perhaps the chorions show adaptations that are present in all species that make a move toward incubation in the intertidal zone fairly straightforward. *C. acuticeps* is a congener of intertidal species such as *C. analis* and *C. recalvus*, and a comparison of their eggs as well as their reproductive behavior and adult physiology could reveal more adaptations that make this early life habitat shift possible.

Among the fundulids, numerous species show semilunar rhythms that indicate tides synchronize spawning (Taylor, 1984; Martin et al., 2004). Examining relationships among 14 species within the Fundulidae, Bernardi (1997) placed *F. heteroclitus* and *F. grandis* within one clade, and five species within this clade show similar reproductive behavior but different choices of spawning substrate (Able & Hata, 1984). However, another beach-spawning fundulid, the diamond killifish *Adinia xenica*, was not placed close to that clade. Many of the fundulid species that show semilunar spawning were not included in this analysis, making it difficult to draw phylogenetic conclusions.

Ghedotti and Davis (2013) provide a more recent analysis, and emphasize salinity tolerance because of the fact that these killifishes are found in many different habitats, from freshwater to marine environments. The ready adaptation to different salinities undoubtedly allows the fishes in this rapidly evolving family to colonize new habitats and spawn on novel substrates. In their analysis, one clade with high salinity tolerance is composed entirely of five species that are known or suspected to spawn on beaches with semilunar rhythmicity. These are *F. jenkinsi*, *F. confluentes*, *F. pulvereus*, *F. grandis*, and *F. heteroclitus*. However, two other species of fundulids that spawn on beaches were placed in two very different branches of the family tree, *F. similus* and the diamond killifish (Ghedotti & Davis, 2013), indicating the possibility of three separate origins of beach spawning within this family. A broad salinity tolerance appears to be the ancestral condition in this group and it is not impossible that beach spawning behavior is also an ancestral character. Additional information about the phylogeny of more species within this clade may clarify this.

1.3   SOME LINEAGES INDICATE INDEPENDENT ORIGINS OF BEACH-SPAWNING BEHAVIOR

A phylogenetic approach considers how and in which groups beach spawning developed. One form of inquiry compares closely related species within groups that contain one or more species of beach-spawning fishes.

The puffer fishes in the genus *Takifugu* (Tetraodontidae) show an explosive species radiation within the last 5 to 10 million years (Kato et al., 2005) that Yamanoue et al. (2008) compare to the cichlid radiations in Malawi and Lake Tanganyika. Most *Takifugu* species inhabit coastal marine waters throughout life, but *T. niphobles* is the only one of 25 closely related species that spawns on beaches (Uno, 1955; Yamahira, 1996). The kusafugu fish gather in huge numbers near a gravel beach until a single female, attended by up to 60 males, approaches the shore and broadcasts her eggs among the stones. For more on *T. niphboles* spawning behavior, see Chapter 4.

Reproductive behavior has not yet been characterized in all the *Takifugu* species. However, of these species, *T. niphobles* is most closely allied to *T. ocellatus*

(Yamanoue et al., 2008). *T. ocellatus* and *T. obscurus* have spawning migrations from seawater to freshwater of the Yangtze River (Yang & Chen, 2008).

T. rubripes, sympatric in Japan with *T. niphobles*, spawns at a depth of 20 m and attaches its eggs to rocks. Mariculture of *T. rubripes* provides food in Japan (Kai et al., 2005). The cultured fish lack the toxic qualities of wild members of this species. Wild *T. rubripes* show site fidelity to natal spawning grounds (Nakajima & Nitta, 2005). Similarly, spawning runs of *T. niphobles* occur on certain beaches with regularity (Yamahira, personal communication), indicating consistency of behavior, but natal site fidelity has not been confirmed for this species.

The 25 genera of toadfishes (Batrachoididae) contain 78 species (Greenfield et al., 2008), but as in Tetraodontidae, only one species is known to spawn on beaches. The plainfin midshipman *Porichthys notatus* lives subtidally but spawns intertidally (Arora, 1948). Each male sets up a territory in a space hollowed out beneath a boulder, where he may enjoy multiple matings followed by nest guarding of the eggs and larvae for additional weeks (Arora, 1948; Crane, 1981). For details of its mating behavior, see Chapter 4.

P. notatus has been a model organism for the study of sound production during reproductive behavior (Brantley & Bass, 1994; Rice & Bass, 2009), but reproduction in its congener, *P. meriaster*, has not been extensively described. There is no published evidence that *P. meriaster* spawns intertidally, although this species otherwise seems to replace *P. notatus* south of Pt. Conception along the California coast. On the other hand, both *P. notatus* and *P. myriaster* are reported to behave similarly during reproduction (Feder et al., 1974). Males of both species produce underwater sounds by humming, vibrating the swim bladder to attract females to nests under rocks. The presence of photophores on the ventral surface of both *Porichthys* species is remarkable for fishes that frequent shallow waters, and both of these species can also be found at depths to 120 m (Eschmeyer et al., 1983).

Females of both species produce eggs that are attached to the underside of a boulder by an adhesive disk that persists after hatching, tethering the young hatchlings and larvae to the boulder during early development (Arora, 1948; Crane, 1981). These odd stalked eggs that hold the young larvae for days or weeks after hatching are seen in other species of Batrachoididae, including *Opsanus tau* in the Atlantic Ocean (Breder & Rosen, 1966). The unique ecology of *P. notatus*, with adult males, developing eggs, and hatchlings all subjected to tidal exposure in air, deserves additional attention from researchers. A more detailed examination of reproduction and early life history in *P. myriaster* may help to elucidate how adaptations to this labor-intensive form of reproduction began.

1.4   SOME TELEOST FISH SPECIES SPAWN ON BEACHES AND OTHER HABITATS

The capelin *Mallotus villosus* (Osmeridae) migrates to shore from deeper water to spawn on sandy or gravel beaches of Canada, Alaska, Iceland, and other circumpolar coasts (Hart, 1973). Spawning takes place over a series of days when high wind waves lift the capelin toward the shore to spawning sites (Figure 1.1). For more about their spawning behavior, see Chapter 4.

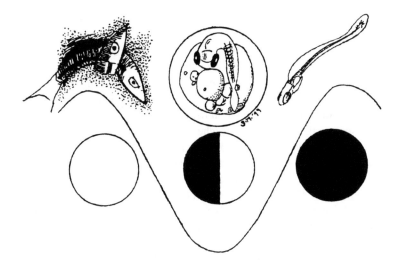

FIGURE 1.1 California grunion *Leuresthes tenuis* spawn on the highest tides of the new or full moon, leaving their eggs high on shore. The embryos hatch as tides rise two weeks later for the next semilunar high tide. (From Martin, K. L. M., *Physiological Zoology*, 69, 1096–1113. Art by Gregory Martin.)

M. villosus is also known to spawn in deeper water on occasion (Frank & Leggett, 1981; Nakashima & Wheeler, 2002). High air temperatures seem to drive this change in spawning location during summer, as there is no genetic differentiation between the populations of capelin spawning in these different habitats (Carscadden et al., 1991). This indicates plasticity in reproductive behavior in response to an environmental cue and may provide flexibility in spawning site choice.

The obvious assumption that beach spawning in this species evolved from subtidal spawning may not be correct in this case, however. Off the coast of Newfoundland, Canada, one subtidal spawning location was identified as being a former beach site during the last glaciation period (Nakashima & Taggart, 2002; Nakashima & Wheeler, 2002). If this species persisted in fidelity to this spawning site over a time of sea level rise, subtidal spawning could have resulted when the beach was drowned. If so, beach spawning was the antecedent to subtidal spawning in this case, instead of the other way around. This may have implications for the future survival of many beach-spawning species, as survival of capelin embryos in these subtidal sediments can be very low (Nakashima & Wheeler, 2002). See Chapter 9 for more on threats to beach-spawning fishes.

Capelin males may only reproduce once, at the age of two or three years. It was previously suggested that this species may pursue a gender-specific life history strategy (Huse, 1998) with males dying after spawning but females spawning more than once. However, it appears that for at least some populations, after spawning, nearly all the adults die (Nakashima & Taggart, 2002). Their bodies may wash up on nearby beaches and provide food to scavengers, connecting these terrestrial birds and mammals to the oceanic food web. Christiansen et al. (2008) found that capelin of both sexes that spawned subtidally all were semelparous and died after spawning.

However, both males and females that spawned on beaches were all iteroparous, able to return and spawn again.

Within the osmerid clade, anadromy is the ancestral condition as many ocean-going species return as adults to swim upstream into freshwater to spawn, similar to many Salmonids. Although there are several beach-spawning osmerids, none of these smelt genera are closely related to *M. villosus* (Martin & Swiderski, 2001), and all have sister taxa that do not spawn on beaches. See Section 1.4.

Another species that spawns on beaches and elsewhere is the threespine stickleback, *Gasterosteus aculeatus* (Gasterosteidae). This species occurs in a wide variety of freshwater, brackish, and seawater habitats (Hart, 1973). This species is well studied for its mating rituals involving construction of a nest that is then guarded by the male. However, some populations spawn without constructing a nest or providing any kind of parental care. Sticklebacks as a whole show extremely rapid evolution and reproductive isolation through changes in behavior (McKinnon & Rundle, 2002; Shaw et al., 2007).

One group within the *G. aculeatus* complex, the white stickleback, is possibly an incipient species. White stickleback populations are distinguished by the bright white coloration of the males and also show changes in breeding behavior. In one location, one group of white *G. aculeatus* spawns over subtidal algae, while another group spawns over bare rock in the intertidal zone and has a complete lack of nest building or parental attendance to the clutch after spawning (MacDonald et al., 1995a,b). See Chapter 4 for more on their spawning behavior.

Both groups of white stickleback are genetically similar to "traditional" threespine stickleback males that reproduce in elaborate nests on vegetated substrates and stay home with their offspring. This indicates that rapid behavioral change has occurred during critical reproductive events in the absence of genetic differentiation (Haglund et al., 1990; Shaw et al., 2007). Comparisons among populations within this species or species complex would be fruitful to consider embryonic acclimation to environmental conditions across these widely different spawning microhabitats.

Similarly to the sticklebacks, other species also have only one population that spawns on beaches. Most salmon are anadromous, but one population in Alaska spawns in the intertidal zone of a beach. Some rock sole occasionally spawn on a beach in Puget Sound (see Chapter 4). This behavior may be more common than is currently appreciated and may turn up when least expected.

1.5   LINEAGES THAT INCLUDE MULTIPLE SPECIES OF BEACH-SPAWNING FISHES SHOW MULTIPLE INDEPENDENT ORIGINS OF THIS BEHAVIOR

Several lineages or families of teleost fishes have multiple species that live in subtidal or estuarine habitats and migrate inshore to spawn on beaches, including Fundulidae, Osmeridae, and Atherinopsidae (Taylor, 1984; Martin, 1999; Martin & Swiderski, 2001). Both Fundulidae and Atherinopsidae are within the Atherinomorpha clade, but diverge at the ordinal level between Cyprinodontiformes and Atheriniformes (Nelson, 2006). Osmerids are in the Osmeriformes, sister order to the Salmoniformes within the Protacanthopterygii clade.

Considering first the Atherinopsidae (Figure 1.2), the group as a whole is considered estuarine in origin (Bamber & Henderson, 1988). Silversides in general show large spawning aggregations and depend on a suitable substrate for oviposition, but otherwise no obvious gradients or generalizations indicate which species are likely to develop beach spawning in this group.

There are several beach-spawning species of Atherinopsidae. Two are found in the genus *Leuresthes*. Both of these spawn during semilunar high tides and leave their clutches of eggs high in the intertidal zone (Walker, 1952). See Chapter 4 for more details. Bernardi et al. (2003) suggest that *L. tenuis* diverged from ancestral *L. sardina* during the rifting event that created the Gulf of California a few million years ago. If so, then beach spawning may have originated in the ancestral population that still spawns in shallow waters at the top of the Sea of Cortez (Thomson & Muench, 1976). However, the sister species to this genus, the jacksmelt *Atherinopsis californiensis*, is not known to spawn on beaches, nor does it synchronize its reproduction with tides.

Another atherinopsid, *Colpichthys regis*, sometimes called the false grunion, also lives in the Gulf of California and spawns on beaches. During morning high tides on two or three days at the highest tides around the new and full moons, the fish approach the water line and spawn in very shallow water (Russell et al., 1987). Spawning depends on the availability of the appropriate substrate, either succulent vegetation around mangrove roots or a tossed-aside pile of tiles with numerous crevices that shelter developing eggs. Because of the low-energy wave environment where this species occurs, spawning takes place in very shallow water at the greatest height of the tide. This behavior is very similar to the spawning behavior seen in *L. sardina* and even *L. tenuis* in the same sort of estuarine or bay habitats (Martin et al., 2013), rather than the full emergence from water seen by *L. tenuis* on the outer coast of California, with much more intense waves.

The developing embryos of *L. tenuis* and *L. sardina* are continuously emerged into air but buried under several centimeters of sand, waiting to hatch at the following

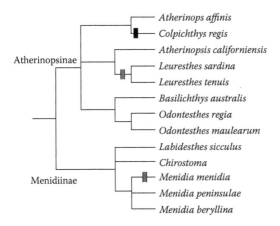

FIGURE 1.2 Cladogram for fishes in the family Atherinopsidae indicating the appearance of beach-spawning behavior. (Based on Martin & Swiderski (2001); cladogram after Crabtree (1987) and Dyer & Chernoff (1996).)

semilunar tides (Moffatt & Thomson, 1978; Griem & Martin, 2000; Martin et al., 2011). Embryos in the eggs of *C. regis* are emerged into air frequently but sheltered by their surrounding substrate to some extent (Russell et al., 1987; D. Middaugh, personal communication). Details of their incubation time are not known, nor is it known whether they hatch in response to an environmental trigger.

Another related species, the Delta silverside *C. hubbsi* (Crabtree, 1989), is found in the Colorado delta at the apex of the Gulf of California, over sandy and muddy substrates in freshwater and estuarine habitats. The reproduction of this endangered species has not been described. The most closely related sister genus to *C. regis* is the topsmelt *Atherinops affinis*, an estuarine species that spawns on vegetation without any apparent tidal synchrony (Crabtree, 1987).

A fourth beach-spawning Atherinopsid species, the Atlantic silverside *Menidia menidia*, is found in estuaries along the North American coast from Quebec, Canada, to northern Florida. Classified within a separate subfamily from the above species (Dyer & Chernoff, 1996), it spawns over seagrass beds within estuaries or bays during high semilunar tides (Middaugh, 1981; Middaugh et al., 1983).

Members of the genus *Menidia* and others in the subfamily Menidiinae have moved into freshwater from an estuarine origin (Bamber & Henderson, 1988). A congener of *M. menidia*, the tidewater silverside *M. peninsulae*, occurs in estuaries along the Gulf of Mexico and southeastern US coast and shows some semilunar periodicity in spawning (Middaugh & Hemmer, 1984). The Mississippi silverside *M. audens* occurs in freshwater but tolerates brackish water and can make short trips through seawater. This species has invaded some estuaries on the California coast from freshwater introductions (C. Swift, personal communication).

Among the osmerids are four species that spawn on beaches, the capelin *M. villosus* discussed above, two species of the day smelt *Hypomesus,* and the night smelt *Spirinchus starksi* (Figure 1.3). All these osmerids spawn on gravel beaches during high tides in such shallow water that the eggs become tidally emerged frequently during incubation (Breder & Rosen, 1966). For details on the spawning behavior see Chapter 4. Their small eggs adhere individually to gravel and resist desiccation and abrasion encased in tough chorions. Both *H. pretiosus* and *S. starksi* may occur on the same beaches on the west coast of North America (Sweetnam et al., 2001).

The Osmeriformes clade groups with the Salmoniformes (Nelson, 2006), and it is likely that the ancestral spawning strategy was anadromy (Martin & Swiderski, 2001). Among the species shown in Figure 1.3, of those that do not spawn on beaches, all but one are anadromous. This includes *Spirinchus thaleichthys*, a congener of *S. starksi*. The North American whitebait smelt, *Allosmerus elongatus*, is the exception that lives and spawns subtidally.

Within the genus *Hypomesus*, two species *H. pretiosus* and *H. japonicus,* are both marine and spawn on beaches in similar fashion (Hirose & Kawaguchi, 1998; Quinn et al., 2012). However their congener, the Delta smelt *H. transpacificus*, occurs in central California and is considered anadromous, spawning in freshwater and migrating into estuaries (Eschmeyer et al., 1983).

Thus, despite the similarities in spawning behaviors among the beach-spawning Osmerids, these three species are only distantly related to one another within

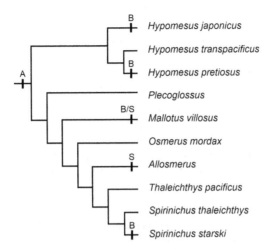

FIGURE 1.3 Cladogram for fishes in the family Osmeridae indicating the appearance of beach-spawning behavior. A is anadromy, the ancestral condition; B indicates beach-spawning behavior; S indicates subtidal spawning. (Based on Martin & Swiderski (2001); cladogram after Johnson & Patterson (1996).)

the clade. The evidence, even within genera, indicates at least three different independent origins of beach spawning from an ancestral condition of anadromy (Martin & Swiderski, 2001).

1.6 THE LEAP OF FAITH, PLASTIC BEHAVIOR, AND EVOLUTION

Antecedent behavior is not consistent between different examples of species pairs and does not predict beach spawning in fishes. In the evolution of beach-spawning behavior, species tend to show relatively plastic behavior during reproduction (Martin & Swiderski, 2001) and possess a physiology that tolerates, at least temporarily, rapid changes in the environment (Hofmann & Todgham, 2010). All beach-spawning fishes reproduce with large, adherent demersal eggs that require long incubation times of many days or even weeks. Regulatory genes assist the survival of eggs to respond rapidly to exposure in air (Tingaud-Sequeira et al., 2013).

Spawning behaviors vary between closely related species and across clades. Possible precursor behaviors might include subtidal spawning, estuarine spawning, intertidal spawning synchronized with high semilunar tides, anadromous migrations, ovipositing on substrates such as vegetation or bare rocks or gravel or sand, parental care or no parental care, and site fidelity or variability of spawning site. Chapters 3 and 4 describe the reproductive behaviors of many of these species.

Even within single lineages, multiple beach-spawning species appear to have had independent origins of this reproductive behavior. It seems to be an adaptation for the benefit of the early life stages, even as it puts the parents in harm's way. This striking leap of faith stands out in diverse species wherever it occurs. Chapter 2 examines some of the physical conditions of these beach habitats and the geographic extent of fishes that spawn on beaches.

REFERENCES

Able, K. W. & Hata, O. (1984). Reproductive behavior in the *Fundulus heteroclitus–F. grandis* complex. *Copeia* 1984, 820–825.

Almada, V. C. & Santos, R. S. (1995). Parental care in the rocky littoral: adaptation and exaptation in Atlantic and Mediterranean blennies. *Reviews in Fish Biology and Fisheries* 5, 23–37.

Arora, H. L. (1948). Observations on the habits and early life history of the Batraochoidid fish, *Porichthys notatus* Girard. *Copeia* 1948, 89–93.

Bamber, T. N. & Henderson, P. A. (1988). Pre-adaptive plasticity in atherinids and the estuarine seat of teleost evolution. *Journal of Fish Biology* 33, 17–23.

Bernardi, G. (1997). Molecular phylogeny of the Fundulidae (Teleostei, Cyrpindontiformes) based on the cytochrome b gene. In *Molecular Systematics of Fishes*, Kocher, T. D. & Stepien, C. A. (eds.), pp. 189–197. Academic Press, San Diego, CA.

Bernardi, G., Findley, L. & Rocha-Olvares, A. (2003). Vicariance and dispersal across Baja California in disjunct marine fish populations. *Evolution* 57, 1599–1609.

Brantley, R. K. & Bass, A. H. (1994). Alternative male spawning tactics and acoustic signals in the plainfin midshipman fish *Porichthys notatus* Girard (Teleostei, Batrachoididdae). *Ethology* 96, 213–232.

Breder, C. M. & Rosen, D. E. (1966). *Modes of Reproduction in Fishes*. The Natural History Press, Garden City, NJ, 941 pp.

Carscadden, J. E., Frank, K. T. & Miller, D. S. (1991). Capelin (*Mallotus villosus*) spawning on the southeast shoal: influence of physical factors past and present. *Canadian Journal of Fisheries and Aquatic Science* 46, 1743–1754.

Christiansen, J., Praebel, K., Siikavuopio, S. I., & Carscadden, J. E. (2008). Facultative semelparity in capelin *Mallotus villosus* (Osmeridae)—an experimental test of a life history phenomenon in a subarctic fish. *Journal of Experimental Marine Biology and Ecology* 360, 47–55. DOI: 10.1016/j.jembe.2008.04.003.

Crabtree, C. B. (1987). Allozyme evidence for the phylogenetic relationship within the silverside subfamily Atherinopsinae. *Copeia* 1987, 860–867.

Crabtree, C. B. (1989). A new silverside of the genus *Colpichthys* (Atheriniformes: Atherinidae) from the Gulf of California, Mexico. *Copeia* 1989, 558–568.

Crane, J. J. (1981). Feeding and growth by the sessile larvae of the teleost *Porichthys notatus*. *Copeia* 1981, 895–897.

DeMartini, E. E. (1999). Intertidal spawning. In *Intertidal fishes: Life in Two Worlds*. Horn, M. H., Martin, K. L. M. & Chotkowski, M. A. (eds.), pp. 143–164. Academic Press, San Diego, CA.

Dyer, B. S. & Chernoff, B. (1996). Phylogenetic relationships among atheriniform fishes (Teleostei: Atherinomorpha). *Zoological Journal of the Linnaean Society* 117, 1–69.

Eschmeyer, W. N., Herald, E. S. & Hammann, H. (1983). *A Field Guide to Pacific Coast Fishes of North America*. Peterson Field Guide Series, Houghton Mifflin, Boston. 336 pp.

Feder, H. M., Turner, C. H. & Limbaugh, C. (1974). *Observations on fishes associated with kelp beds in southern California*. Fish Bulletin 160, State of California Resources Agency, Department of Fish & Game. 144 pp.

Frank, K. T. & Leggett, W. C. (1981). Prediction of egg development and mortality rates in capelin (*Mallotus villosus*) from meteorological, hydrographic, and biological factors. *Canadian Journal of Fisheries and Aquatic Science* 38, 1327–1338.

Froese, R. & Pauly, D. (eds). (2013). FishBase.org. World Wide Web electronic publication.

Ghedotti, M. & Davis, M. (2013). Phylogeny, classification, and evolution of salinity tolerance of the North American topminnows and killifishes, family Fundulidae (Teleostei: Cyprinodontiformes). *Fieldiana Life and Earth Sciences* 7, 1–65. DOI: http://dx.doi.org/10.3158/2158-5520-12.7.1.

Greenfield, D. W., Winterbottom, R. & Collette, B. B. (2008). Review of the toadfish genera (Teleostei: Batrachoididae). *Proceedings of the California Academy of Sciences*, Fourth Series, 59(15), 665–710.

Griem, J. N. & Martin, K. L. M. (2000). Wave action: the environmental trigger for hatching in the California grunion *Leuresthes tenius* (Teleostei: Atherinopsidae). *Marine Biology* 137, 177–181.

Haglund, T. R., Buth, D. G. & Blouw, D. M. (1990). Allozyme variation and the recognition of the "white stickleback." *Biochemical Systematics and Ecology* 18, 559–563. http://dx.doi.org/10.1016/0305-978(90)90129-4.

Hart, J. L. (1973). Pacific fishes of Canada. *Fisheries Research Board of Canada, Bulletin* 180, 1–740.

Hirose, T. & Kawaguchi, K. (1998). Spawning ecology of Japanese surf smelt, *Hypomesus pretiosus japonicas* (Osmeridae), in Otsuchi Bay, northeastern Japan. *Environmental Biology of Fishes* 52, 213–223.

Hofmann, G. E. & Todgham, A. E. (2010). Living in the now: physiological mechanisms to tolerate a rapidly changing environment. *Annual Reviews in Physiology* 72, 127–145. DOI: 10.1146/annurev-physiol-021909-135900.

Horn, M. H., Martin, K. L. M. & Chotkowski, M. A. (1999). *Intertidal Fishes: Life in Two Worlds*. San Diego: Academic Press, 399 pp.

Ilves, K. L. & Taylor, E. B. (2007). Evolutionary and biogeographical patterns within the smelt genus *Hypomesus* in the north Pacific Ocean. *Journal of Biogeography* 35, 48–64. DOI: 10.1111/j.1365-2699.2007.01782.x.

Johnson, G. D. & Patterson, C. (1996). Relationships of lower Euteleostean fishes. In *Interrelationships of Fishes,* Stiassny, M. L. J., Parenti, L. R. & Johnson, G. D. (eds.), pp. 251–332. Academic Press, San Diego, CA.

Kai, W., Kikuchi, K., Fujita, M., Suetake, H., Fujiwara, A., Yoshiura, Y., Ototake, M., Venkatesh, B., Miyaki, K. & Suzuki, Y. (2005). A genetic linkage map for the tiger puff-erfish, *Takifugu rubripes. Genetics* 171, 227–238. DOI: 10.1534/genetics.105.042051.

Kato, A., Doi, H., Nakada, T., Sakai, H. & Hirose, S. (2005). *Takifugu obscurus* is a euryhaline fugu speices very close to *Takifugu rubripes* and suitable for studying osmoregulation. *BMC Physiology* 5, 18. DOI: 10.1186/1472-6793-5-18.

MacDonald, J. F., Bekkers, J., MacIsaac, S. M. & Blouw, D. M. (1995a). Intertidal breeding and aerial development of embryos of a stickleback fish (*Gasterosteus*). *Behaviour* 132, 1183–1206. http://www.jstor.org/stable/453531.

MacDonald, J. F., MacIsaac, S. M., Bekkers, J. & Blouw, D. M. (1995b). Experiments on embryo survivorship, habitat selection, and competitive ability of a stickleback fish (*Gasterosteus*) which nests in the rocky intertidal zone. *Behaviour* 132, 1207–1221. http://www.jstor.org/stable/453533.

Marliave, J. B. (1981). High intertidal spawning under rockweed, *Fucus distichus*, by the sharpnose sculpin, *Clinocottus acuticeps. Canadian Journal of Zoology* 59, 1122–1125.

Martin, K., Bailey, K., Moravek, C. & Carlson, K. (2011). Taking the plunge: California grunion embryos emerge rapidly with environmentally cued hatching. *Integrative and Comparative Biology* 51, 26–37. DOI: 10.1093/icb/icr037.

Martin, K. L. M. (1996). An ecological gradient in air-breathing ability among marine cottid fishes. *Physiological Zoology* 69, 1096–1113.

Martin, K. L. M. (1999). Ready and waiting: delayed hatching and extended incubation of anamniotic vertebrate terrestrial eggs. *American Zoologist* 39, 279–288.

Martin, K. L. M. & Bridges, C. R. (1999). Respiration in water and air. In *Intertidal Fishes: Life in Two Worlds*, Horn, M. H., Martin, K. L. M. & Chotkowski, M. A. (eds.), pp. 54–78. San Diego: Academic Press.

Martin, K. L. M. & Swiderski, D. L. (2001). Beach spawning in fishes: phylogenetic tests of hypotheses. *American Zoologist* 41, 526–537.

Martin, K. L. M., Heib, K. A. & Roberts, D. A. (2013). A southern California icon surfs north: local ecotype of California grunion, *Leuresthes tenuis* (Atherinopsidae) revealed by multiple approaches during temporary habitat expansion into San Francisco Bay. *Copeia* 2013, 729–739. DOI: 10.1643/CI-13-036.

Martin, K. L. M., Van Winkle, R. C., Drais, J. E. & Lakisic, H. (2004). Beach-spawning fishes, terrestrial eggs, and air breathing. *Physiological and Biochemical Zoology* 77, 750–759.

McKinnon, J. S. & Rundle, H. D. (2002). Speciation in nature: the threespine stickleback model systems. *Trends in Ecology and Evolution* 17, 480–488.

Middaugh, D. P. (1981). Reproductive ecology and spawning periodicity of the Atlantic silverside, *Menidia menidia* (Pisces: Atherinidae). *Copeia* 1981, 766–776.

Middaugh, D. P. & Hemmer, M. J. (1984). Spawning of the tidewater silverside, *Menidia peninsulae* (Goode and Bean), in response to tidal and lighting schedules in the laboratory. *Estuaries* 7, 139–148.

Middaugh, D. P., Kohl, H. W. & Burnett, L. E. (1983). Concurrent measurement of intertidal environmental variables and embryo survival for the California grunion, *Leuresthes tenuis*, and Atlantic silverside, *Menidia menidia* (Pisces: Atherinidae). *California Fish & Game* 69, 89–96.

Moffatt, N. M. & Thomson, D. A. (1978). Tidal influence on the evolution of egg size in the grunions (Leuresthes, Atherinidae). *Environmental Biology of Fishes* 3, 267–273.

Nakajima, H. & Nitta, A. (2005). Homing behavior of adult ocellate puffer *Takifugu rubripes* to the natal spawning ground at the mouth of Isle Bay based on tagging experiments. *Nippon Suisan Gakkaishi* 71, 736–745 (in Japanese, English abstract).

Nakashima, B. S. & Taggart, C. T. (2002). Is beach-spawning success for capelin, *Mallotus villosus* (Muller), a function of the beach? *ICES Journal of Marine Science* 59, 897–908.

Nakashima, B. S. & Wheeler, J. P. (2002). Capelin (*Mallotus villosus*) spawning behavior in Newfoundland waters: the interaction between beach and demersal spawning. *ICES Journal of Marine Science* 59, 909–916.

Nelson, J. S. (2006). *Fishes of the World, 4th Edition.* John Wiley & Sons, Hoboken, NJ, USA.

Petersen, C. W. & Hess, H. C. (2011). Evolution of parental behavior, egg size, and egg mass structure in sculpins. In *Adaptation and Evolution in Cottid Fishes,* Gogo, A., Munehara, H. & Yabe, M. (eds), pp. 194–203. Tokai University Press, Kanagawa, Japan.

Quinn, T., Krueger, K., Pierce, K., Penttila, D., Perry, K., Hicks, T. & Lowry, D. (2012). Patterns of surf smelt, *Hypomesus pretiosus*, intertidal spawning habitat use in Puget Sound, Washington State. *Estuaries and Coasts.* DOI: 10.1007/s12237-012-9511-1.

Rice, A. N. & Bass, A. H. (2009). Novel vocal repertoire and paired swimbladders of the three-spined toadfish, *Batrachomoeus trispinosus*: insights into the diversity of the Batrachoididae. *Journal of Experimental Biology* 212, 1377–1391. DOI: 10.1242/jeb.028506.

Richards, J. G. (2011). Physiological, behavioral and biochemical adaptations of intertidal fishes to hypoxia. *Journal of Experimental Biology* 214, 191–199. DOI: 10.1242/jeb.047951.

Russell, G. A., Middaugh, D. P. & Hemmer, M. J. (1987). Reproductive rhythmicity of the atherinid fish, *Colpichthys regis*, from Estero del Soldado, Sonora, Mexico. *California Fish & Game* 73, 169–174.

Shaw, K., Scotti, M. & Foster, S. (2007). Ancestral plasticity and the evolutionary diversification of courtship behavior in threespine sticklebacks. *Animal Behaviour* 73, 415–422.

Sweetnam, D. A., Baxter, R. D. & Moyle, P. B. (2001). True smelts. In *California's Living Marine Resources: A Status Report,* pp. 470–478. California Department of Fish and Game, Sacramento, CA.

Taylor, M. H. (1984). Lunar synchronization of fish reproduction. *Transactions of the American Fisheries Society* 113, 484–493.

Thomson, D. A. & Muench, K. A. (1976). Influence of tides and waves on the spawning behavior of the Gulf of California grunion, *Leuresthes sardina* (Jenkins & Evermann). *Bulletin of the Southern California Academy of Science* 75, 198–203.

Tingaud-Sequeira, A., Lozano, J.-J., Zapater, C., Otero, D., Kube, M., Reinherdt, R. & Cerda, J. (2013). A rapid transcriptome response is associated with desiccation resistance in aerially-exposed killifish embryos. *PLOS One* 8, e64410.

Uno, Y. (1955). Spawning habit and early development of a puffer, *Fugu (Torafugu) niphobles* (Jordan and Snyder). *Journal of Tokyo University Fisheries* 41, 169–183.

Walker, B. W. (1952). A guide to the grunion. *California Fish and Game* 38, 409–420.

Yamahira, K. (1996). The role of intertidal egg deposition on survival of the puffer, *Takifugu niphobles* (Jordan et Snyder), embryos. *Journal of Experimental Marine Biology and Ecology* 198, 291–306.

Yamanoue, Y., Miya, M., Matsuura, K., Miyazawa, S., Tsukamoto, N., Doi, H., Takahashi, H., Mabuchi, H., Nishida, M. & Sakai, H. (2008). Explosive speciation of *Takifugu*: another use of Fugu as a model system for evolutionary biology. *Molecular Biology and Evolution* 26, 623–629.

Yang, Z. & Chen, Y. F. (2008). Differences in reproductive strategies between obscure puffer *Takifugu obscurus* and ocellated puffer *Takifugu ocellatus* during their spawning migration. *Journal of Applied Ichthyology* 24, 569–573. DOI: 10.1111/j.1439-0426.2008.01071.x.

2 Surf, Sand, and Beach: Physical Conditions of Shore Habitats for Fishes

Beach spawning has evolved multiple times in diverse teleost lineages. A beachfront home provides many benefits to embryos, along with some obvious potential disadvantages. Advantages to the embryos include the potential for increased oxygen diffusion rates and higher temperatures during air exposure, both of which may speed embryonic development, and avoidance of aquatic predators (Taylor et al., 1977; Middaugh et al., 1983; Martin et al., 2009). Disadvantages on the other hand are present for both the embryos and the spawning adults, and include a higher risk of desiccation while in air and exposure to novel predators or pathogens from the land (Martin, 1999; Martin & Strathmann, 1999; Strathmann & Hess, 1999). There is also the chance a larva may hatch into a terrestrial environment for which, as a fish, it is completely unprepared (Frank & Leggett, 1981). Chapter 7 provides more details on the effects of incubation on the beach for the early life stages.

The advantages of spawning on shore outweigh the disadvantages sufficiently enough that even some species of teleosts that usually live in deeper water transport themselves to the beach solely for reproduction (DeMartini, 1999; Martin et al., 2004). This chapter examines the characteristics of the habitats that are conducive to the evolution of beach spawning in fishes.

2.1 TIDAL EBB AND FLOW ALTER HABITAT CONDITIONS RAPIDLY AND PREDICTABLY ON BEACHES

Ocean coastal habitats are subject to tidal ebb and flow. Terrestrial and marine influences interact in the narrow ocean coastal margin (Stevenson & Stevenson, 1972). Between the lowest and highest tides, the intertidal zone is subject to regular cycles of inundation and exposure. In rock pools during low tides, changes may occur in salinity with rainfall or evaporation, or in temperature with insolation or cold night air, or in dissolved oxygen and carbon dioxide as animals and plants respire and photosynthesize. Because of the small volume and temporary lack of exchange with the wider ocean, these changes are far more rapid and extreme than the fluctuations that occur in the open ocean (Truchot & Duhamel-Jouve, 1980). Pools provide aquatic refuge, but large expanses of substrate in the intertidal zone also are exposed to air for hours during a low tide, along with any animals or plants that are attached to or burrowed beneath these surfaces.

With dewatering during low tides and pounding surf during high tides, it may seem that few species would make their homes here, let alone reproduce. But the inputs of nutrients and energy from sea, sky, and land that create such variability in the intertidal zone also make it one of the most productive habitats on Earth (Stevenson & Stevenson 1972; Leigh et al., 1987). Many species of fishes take advantage of the opportunities for feeding, avoiding predators, mating, and nesting in this life-filled zone (Horn & Gibson, 1988; Bridges, 1993; Horn et al., 1999; Horn & Martin, 2006).

The progression of moon phases creates tidal oscillations of variable height, so that on a rocky shelf of bedrock or sandy beach, intertidal plants and animals high on shore are exposed to air much more frequently and for longer periods of time than plants and animals in the mid-intertidal zone (Figure 2.1). Tides oscillate to

(a)

(b)

FIGURE 2.1 (See color insert.)   Vertical height in the intertidal zone is correlated with duration of air emergence. Organisms found higher in the intertidal zone will be emerged for longer periods of time and more frequently than organisms found lower in the intertidal zone. (a) Big Rock Beach at high tide. (b) The same beach at low tide.

a greater extent around the times of the new and full moons, the syzygy tides that occur approximately every two weeks. Annual cycles superimpose on the semilunar and monthly lunar cycles so that the most extreme astronomical tides occur in midwinter and midsummer. Plants and animals in the lowest part of the intertidal zone may be exposed to air only briefly during a few tides of the year, whereas those that live in the highest zones may spend much more time out of water than in it. These reliable microhabitat exposures result in vertical zonation of different species of invertebrates, plants, and fishes at different heights on shore (Stevenson & Stevenson, 1972; Benson, 2002).

2.2 SOME FISH SPECIES ARE RESIDENT IN THE INTERTIDAL ZONES DURING ALL PARTS OF THE TIDAL CYCLE, OTHERS VISIT ONLY AT HIGH TIDE

Some teleost species of fishes that are resident within the intertidal zone tend to stay in place during low tides, coping with all the changes that come to the habitat by air exposure or isolation in pools (Yoshiyama et al., 1992; Pfister, 1992; Horn & Martin, 2006). Many species of Cottidae, Stichaeidae, Clinidae, Gobeisocidae, Blenniidae, Gobiidae, Tripterygiidae, and other teleosts are resident in the rocky intertidal habitat (Horn & Gibson, 1988; Horn et al., 1999).

Among these, more than 60 species of teleost fish from multiple lineages not only emerge from water but also breathe air as tides recede (Bridges, 1988; Sayer & Davenport, 1991; Graham, 1997; Martin & Bridges, 1999; Martin, 2013). For most resident intertidal fishes, exposure to air and terrestrial conditions occurs mainly during low tides. Even these teleosts that reside in the intertidal zone have different tolerances for tidal air emergence (Yoshiyama et al., 1995; Martin, 1996; Sloman et al., 2008), and this influences microhabitat choice.

Fishes are highly mobile, but resident intertidal species are able to navigate home and show fidelity to habitats within specific tidal heights and particular pools (Williams, 1957; Richkus, 1978; Barton, 1982; Yoshiyama et al., 1992; Gibson, 1999). Selection of habitat may differ between sexes (Williams, 1954), with temperature (Nakamura, 1976; Nakano & Iwama, 2002), or by season (Moring, 1986; Yoshiyama et al., 1986; Davis, 2000). Tide pool sculpins that reside in pools in the low intertidal are more likely to be site faithful than residents of mid- and high-intertidal pools (Fangue et al., 2011).

Some teleost species, for example flatfishes, are transient visitors that appear in the intertidal zone only during high tides and move out to sea as the tide falls, when the area is exposed to air (Gibson, 1982; Yoshiyama et al., 1986; Horn & Gibson, 1988; Bridges, 1993; Davis, 2001).

2.3 BEACH SPAWNING FOR FISHES USUALLY OCCURS DURING HIGH TIDES

Beach spawning by fishes in the intertidal zone, whether residents or transients, typically is synchronized around the highest tides. Spawning and nesting around high tide allows fishes to place their clutches vertically on shore to permit some tidal exposure of the developing embryos to air, and the adult fishes remain aquatic during spawning (Jones, 1972; Taylor et al., 1977; Tewksbury & Conover, 1987; Martin et al., 2004).

The tidal height of the spawning run determines the vertical height of oviposition. Placement of eggs in the intertidal zone at high tide near shore ensures that the embryos will be emerged into air by the ebbing tide, either occasionally or continuously (DeMartini, 1999; Martin & Swiderski, 2001; Martin et al., 2004). In the case of capelin, wind waves rather than tides determine the timing of the run and the shore height of oviposition (Frank & Leggett, 1981).

Nests for resident teleosts occur at species-specific tidal heights. When incubation is complete and the embryos hatch, the small larvae and juveniles may occupy pools higher on shore than the adults of the same species (Zander et al., 1999), possibly as a means of avoiding predation or competition by conspecifics.

2.4 BEACH-SPAWNING BEHAVIOR IN FISHES MAY NOT INVOLVE AIR BREATHING

For most species, the adults that engage in beach spawning do not emerge from water, or emerge only partially and accidentally during the spawning run. However, if they do emerge, adults of some beach-spawning species, for example the estuarine mummichog *Fundulus heteroclitus*, have the ability to breathe air (Halpin & Martin, 1999). Others, including the kusafugu puffer *Takifugu niphobles* (Yamahira, 1996) and the California grunion *Leuresthes tenuis* (Martin et al., 2004), apparently do not breathe air. See Chapter 5 for more about air breathing in beach-spawning fishes.

Gas exchange in air is clearly necessary and important for the incubating clutches of beach-spawning fishes when they are exposed by a low tide. The effects and adaptations of oviposition high in the intertidal zone and air exposure on the developing embryos are explored in detail in Chapter 7.

In the past, many evolutionary biologists did not consider adaptation to survive low aquatic oxygen for most marine organisms, because as a rule, in most of the ocean, dissolved oxygen is relatively high and constant (Graham et al., 1978). However, declines in benthic oxygen levels along coasts or in estuaries from eutrophication occasionally create temporarily hypoxic or even anoxic "dead zones" that may suffocate and kill most living creatures present (Martinez et al., 2006; Diaz & Rosenberg, 2008). Temporary, catastrophic hypoxia can have devastating effects on populations of fishes, but these dead zones far below the surface do not seem to occur in such a way as to select for adaptations for air breathing, although they may facilitate selection for hypoxia tolerance.

Many paleontologists in the past, including Inger (1957) and Romer (1967), suggested the idea that vertebrates emerged onto land from freshwater, and this was promoted by neontologists as well (Randall et al., 1981; Janis & Farmer, 1999). This hypothesis developed from observation that many species of air-breathing aquatic fishes can be found today in swamps and stagnant freshwater. Recently some paleontologists have suggested instead that terrestrial vertebrates arose from the sea, and in particular that they may have arisen from the intertidal zone (Schultze, 1999; Clack, 2007; Niedzwiedzki et al., 2010). Indirect support for this hypothesis comes from intertidal fishes (Graham & Lee, 2004).

Hypoxic conditions occur with predictable regularity on a small scale within the marine intertidal zone. The ebb tides daily expose some portion of the coastal edge into air, and the retreat of water also isolates small pools from the open ocean, providing refuge to aquatic animals. In this constantly changing tide pool environment, conditions may depart greatly from the surrounding ocean. Pools in the intertidal zone may be well oxygenated during high tides because of wave action and connection with the wider ocean. At low tides, during the day tide pools may maintain high oxygen tensions with photosynthesis, but at night, oxygen levels may decline in crowded pools because of still water and the respiration of all resident animals and plants (Truchot & Duhamel-Jouve, 1980).

Crevices and basins in rocky habitats where seawater collects during low tides offer a refuge from air exposure for mobile marine intertidal animals. The timing of the low tide affects the dissolved aquatic oxygen for marine organisms concentrated in these pools (Truchot & Duhamel-Jouve, 1980). During the day, photosynthesis by plants keeps oxygen levels high, but during nocturnal low tides, pools separated from the open ocean may become hypoxic, hypercarbic, and more acidic due to respiration of the many denizens taking refuge within the pool, coupled with the nocturnal lack of photosynthesis.

To avoid the potential problem of embryos being exposed to aquatic hypoxia during low tides, some intertidal species glue their clutches to the walls or ceiling of the incubation chamber, on the bottom of a boulder that shelters a shallow pool at its base (Crane, 1981; Coleman, 1999; Shimizu et al., 2006). As the tide recedes, the small amount of water that remains can provide humidity but may not be in direct contact with the developing embryos or their eggs.

2.5 BEACH SPAWNING OCCURS ON SPECIFIC SUBSTRATES

Habitats in the marine intertidal zone are classified according to the substrate of the ocean floor. This affects the types of organisms that can live there. Many teleost species have found ways to nest in these coastal locations, including gravel beaches (Frank & Leggett, 1981; Yamahira, 1996), sandy shores (Walker, 1952; Thomson & Muench, 1976), estuaries (Taylor, 1999), tide pools (Marliave & DeMartini, 1977), and mud flats (Swenson, 1997; Ishimatsu et al., 2007).

Substrates of rock, sand, gravel, or mud have very different features for intertidal animals and plants, and all have their own types of communities. Substrates on beaches are the result of geologic history, erosion, and wave climate (McLachlan & Brown, 2006). Many areas have mixed habitats. As a general rule, lower wave energy striking the shore leads to the smaller particles making up a beach. Thus, headlands where waves converge tend to have rocky shelves at the base, while quiet estuaries tend to have muddy floors.

Sand, gravel, and cobble beaches develop when appropriate substrates erode and arrive from rivers, bluffs, and longshore transport (Griggs et al., 2005). Each type of sediment has its own characteristics that affect its use by fishes for oviposition. Sand and gravel are porous and hold humidity well while allowing rapid diffusion of oxygen through air spaces. Mud becomes anoxic within a few centimeters of the surface, and its tiny interstitial spaces allow no diffusion of oxygen from surrounding waters. Rocks may hold liquid water as pools, but this water can become hypoxic if stagnant during a low tide.

The two important but opposing needs of embryos in the intertidal zone are to prevent desiccation and to avoid hypoxia. Gravel and sand beaches are both made up of coarse, unconsolidated sediments, so burrows are temporary and animals that live in the interstitial spaces are mobile (McLachlan & Brown, 2006). When tides recede, pools rarely form; but if shallow pools form, they contain no seaweed or attached vegetation and do not provide much protection from avian or terrestrial predators for those animals within them.

On sandy beaches, beach-spawning fishes bury their eggs under a few inches of sand, allowing them to develop in a moist microhabitat that has less-extreme temperatures than the sand surface. Fishes that spawn on gravel habitats may have adhesive eggs that attach individually to pieces of gravel and may be buried by the spawning activity or by wave action after the adults have left. Gravel beaches located in the cloudy, damp Puget Sound house several species that spawn on beaches. Their adhesive eggs survive aerial emergence even at the surface or with only shallow burial, but shaded eggs survive better than those exposed to the full force of the sun when sunny days prevail (Penttila, 2001; Rice, 2006).

Muddy intertidal habitats are likely to be estuarine or in sheltered embayments. These quiet habitats are full of nutrients and free from some of the larger marine predators, but few species of fishes are able to live within estuarine waters because of their inherent variability. Fishes that live in estuaries must tolerate fluctuating conditions including daily changes in salinity, current direction, oxygen tensions, temperatures, and water depth. Eggs in these habitats must be able to do the same, or avoid at least some of these by being emerged into air during low tides, when hypoxia and hypercarbia are most likely to occur. Beach-spawning estuarine fishes may attach their eggs to vegetation such as surfgrass (Middaugh, 1981) or deposit the eggs in mussel shells (Taylor, 1999). Not all estuarine fishes spawn on beaches; some release their eggs into the plankton and others attach theirs to vegetation in deeper water.

Burrows built by fishes and invertebrates in the muddy intertidal zone may be semipermanent. These burrows may be used as refuges for adult fishes and invertebrates during low tides, even though the water within becomes very hypoxic over time (Ishimatsu & Gonzales, 2011). Some species of mudskippers lay eggs within burrows. To avoid aquatic hypoxia for the embryos, the parents bring mouthfuls of air to fill the nest chamber during the incubation period (Ishimatsu & Graham, 2011). See Chapter 3 for more on this species.

The mangrove killifish *Kryptolebius marmoratus* has not been observed spawning in the wild, but its eggs have been found out of water, protected by leaf litter (Abel et al., 1987). Adults of this species are able to spend weeks out of water, hidden in rotted logs (Taylor et al., 2008). Although this species is a model organism in the laboratory, relatively little is known about its natural reproduction (Taylor, 2012), although studies are under way.

2.6 FEW SPECIES OF FRESHWATER FISHES SPAWN AT THE WATER'S EDGE

A few species spawn above the water's edge in freshwater environments, under special circumstances. Freshwater habitats are replete with more than 13,000 fish species (Cohen, 1970). Of these, only a very few have been described as placing their nests terrestrially or on beaches. There undoubtedly are others that have not yet been described, waiting for an observant field biologist. To emerge eggs into air, fishes may depend on evaporation as a mechanism rather than tides. Oviposition at the water's edge or in temporary waters may cause the clutch to be emerged into air at some point.

One type of habitat where freshwater fishes produce terrestrial eggs is the ephemeral pools found in South America and Africa. Annual killifishes such as *Austrofundulus limnaeus* spend their brief lives in temporary ponds that seasonally fill, then dry out. The eggs they produce are so resilient that this life cycle stage enables this species to endure in a dry, anoxic habitat that would otherwise be inhospitable for fishes (Podrabsky et al., 2001). Eggs buried in the mud at the pool's edge survive the annual drought conditions by becoming dormant for an extended period, then resuming development after a complex series of metabolic changes (Podrabsky et al., 2007, 2010; Meller et al., 2014). Although not the focus of this book, the biochemistry of embryonic diapause is a fascinating solution to the problem of fishes overwintering in a dry habitat. See Chapter 7 for more on this adaptation.

Another freshwater habitat with terrestrially spawning fishes, also in South America, is found in slow-moving tributaries of the Amazon River. Leafy vegetation overhanging the river is the substrate for eggs deposited by *Copella arnoldi* (Characidae), presumably to avoid aquatic predation (Breder & Rosen, 1966). The male parent keeps the clutch damp by splashing it every few minutes during the incubation period (Krekorian, 1976). These fish are called splash tetras when cultured for home aquariums, for obvious reasons.

In Central America in Panama, another freshwater characid, *Brycon petrosus*, can move between aquatic habitats over land (Kramer, 1978). Individuals were observed that appeared ripe and ready to spawn, but no nests were found either in the water or terrestrially at that time, and natural reproduction in this species has not been reported since.

The walking catfish *Clarias batrachus* is a rare example of a freshwater amphibious fish; the majority of amphibious fishes are found in the marine intertidal zone (Graham, 1976; Horn et al., 1999). This invasive species lives in many water bodies in the United States (Courtenay et al., 1974) and makes mass migrations during rainy periods across damp ground and in very shallow water in order to spawn. The parents make nest hollows in ephemeral ponds during flooding (Hensley & Courtenay, 1980). Both parents protect the nest during the 30-hour incubation. This use of temporary pools for reproduction greatly expands the habitat available for nesting, as long as some connection to more permanent aquatic habitat remains.

For freshwater fishes, the few terrestrially nesting species are all found in the tropics, where high heat decreases aquatic oxygen solubility. Freshwater ponds that seasonally dry provide a different kind of interface between water and land from the tidal excursions on marine shores. Creek banks with overhanging vegetation provide yet another for *Copella*. Terrestrially nesting fishes are known from a few locations in South America and Africa and probably occur in additional sites. Many fishes in the families Characidae, Gobiidae, and Rivulidae have not been studied for their reproductive behavior, with potential to reveal additional terrestrially spawning species.

Many tropical amphibians are known to nest on vegetation over water or in foam nests that float at the surface or are completely terrestrial (Thibaudeau & Altig, 1999; Summers et al., 2006; Martin & Carter, 2013). Although some frog nests produce eggs with direct development into tiny adults, most of these embryos hatch into aquatic larvae. The widespread implementation of terrestrial nesting among amphibians, in contrast with its rarity among freshwater fishes, is intriguing. Amphibian eggs tend to be larger than teleost eggs. Amphibian clutches tend to be held together with gel, whereas teleost eggs may adhere to one another but it is unusual to find any gel or other matrix surrounding them. The gel helps prevent desiccation and may be replenished with water by a parent during incubation. In addition, the fact that adult amphibians typically are terrestrial, but adult fish are not, makes terrestrial nesting much less problematic for the amphibians and provides perhaps an added impetus to bypass the aquatic larval stage completely.

It is intriguing to consider the number of fishes found in aquatic tropical ecosystems that would occasionally be exposed to aquatic hypoxia. The evolution of air breathing is relatively common in these environments, yet the evolution of terrestrial nesting is extremely rare—or perhaps more common than currently recognized as it is rarely observed. One possibility is that the tropical amphibians have already taken that role; another is that the seasonal reproduction of most species already takes advantage of favorable conditions for teleost embryos. Additional behavioral and metabolic adaptations for spawning in areas of stagnant freshwater are certain to be found in the future.

2.7 AIR, FRESHWATER, AND SEAWATER HAVE VERY DIFFERENT PROPERTIES AS RESPIRATORY MEDIA

Air-breathing fishes are found in habitats that regularly undergo hypoxia; therefore, these fishes tend to live in shallow water or at the water's edge (Graham, 1997). Fishes that emerge from water and spend time on land are considered amphibious, and most of these also breathe air. However, not all air-breathing fishes are amphibious, and not all air-breathing or amphibious fishes spawn on beaches. Beach-spawning fishes do not all breathe air, but they may be found in similar habitats as those that select for air breathing in fishes. See Chapter 5 for more details.

Aerobic respiration is a metabolic process requiring oxygen to release energy from organic chemicals for use in the body. The final products of the process are water and carbon dioxide. The exchange of oxygen and carbon dioxide for respiration is vital to vertebrate life, whether this exchange between the blood and the medium takes place in seawater, air, or freshwater.

For respiratory gas exchange, both oxygen and carbon dioxide move by diffusion based on differences or gradients of partial pressures. Each gas contributes a percentage of the total atmospheric pressure. In Earth's atmosphere, the air is about 79 percent nitrogen and 21 percent oxygen, with small but significant amounts of carbon dioxide and other gases. As the atmosphere is gaseous, the volume of oxygen is also 21 percent of the volume, or 210 mm of oxygen in 1 liter of air. Water that is in equilibrium with the atmosphere will show the same partial pressure of oxygen, but because the oxygen must be dissolved, its volume at equilibrium also depends on the solubility of that gas in water, or whatever solution is under consideration.

Temperature and salinity affect the solubility of respiratory gases in water, and oxygen is far more soluble in fats than in water. Carbon dioxide is more soluble in water than oxygen, because it can combine with water to form the weak carbonic acid. Higher salinities reduce solubility of oxygen in water, as do higher temperatures. Thus warm seawater holds less oxygen per unit volume than cool freshwater, even when both are equilibrated to 21 percent oxygen in the atmosphere. Both are far lower in oxygen per unit volume than air. If we consider 21 percent in parts per million, oxygen in air by volume is 210,000 ppm. In freshwater, oxygen dissolved at the atmospheric partial pressure of 21 percent is around 8.8 ppm at 22°C. Seawater is even lower (Table 2.1). This means that for a fish to obtain a small volume of oxygen, a large volume of water must move across the respiratory membranes. No wonder so many fish species breathe air!

Water is more viscous than air as a respiratory medium; this is obvious when one considers the pressure against an arm as it is swept around in the air, then through the same distance of water. Fishes breathing aquatically, as they do, must move large volumes of water continuously over the gills in order to extract the small amounts of oxygen it holds. Even though fishes in general have low metabolic rates compared with similarly sized mammals, they tend to have very high rates of ventilation by the gills to keep that oxygen moving into their bodies. Because of its high solubility in water, the resultant carbon dioxide easily escapes across the gills following its partial pressure gradient under most conditions.

Surface waters of lakes and oceans tend to be in equilibrium with the atmosphere because of diffusion and mixing by waves and wind. However, aquatic hypoxia may

---

TABLE 2.1

Dissolved Oxygen for Freshwater and Ocean Water at Four Temperatures Shows That as Salinity and Temperature Increase, Oxygen Solubility Decreases—Warm Saltwater Holds Far Less Oxygen per Volume Than Cold Freshwater When Both Are in Equilibrium with Air

	10°C	20°C	30°C	40°C
Freshwater	11.3 mg/L	9.1 mg/L	7.5 mg/L	6.4 mg/L
Seawater	8.8 mg/L	7.2 mg/L	6.1 mg/L	5.3 mg/L

Note: Air is 21% oxygen, or 21 parts in 100; the unit mg/L is equivalent to parts per million.

occur in tide pools separated from the open ocean, especially at night as described above, or in stagnant pools such as freshwater swamps or brackish estuaries, because of oxygen demand from animals, respiring plants, and microbes (Truchot & Duhamel-Jouve, 1980). In still water, diffusion of oxygen is quite slow, even in the presence of a strong partial pressure gradient. Thus fishes in hypoxic habitats must respond with either increased ventilation rates, by moving even higher volumes of water across their gills, or by reducing energy use and tolerating hypoxia, or by finding a new source of oxygen, atmospheric air. Diffusion of oxygen in air is rapid, and aerial hypoxia occurs only rarely under special conditions of enclosure such as burrows or caves, or at high altitudes.

Embryos within eggs cannot break up boundary layers of stagnant water by moving, nor can embryos move their eggs away from hypoxic conditions. Yet embryonic development requires aerobic metabolism, and fishes in general develop rapidly. Thus, beach spawning may be considered a means for embryos to avoid aquatic hypoxia because it allows the eggs and embryos to emerge into air predictably, periodically. Thus the parental oviposition supports survival of the embryo by providing appropriate incubation conditions.

2.8   BEACH-SPAWNING FISHES ARE GLOBAL IN DISTRIBUTION

By its nature, beach spawning depends on the interface between water and land. This geography is typically the coastal marine intertidal zone or the edge of an estuary. Fishes that spawn on coastal marine beaches occur from tropical and temperate to circumpolar Arctic beaches, on at least six of the seven continents, including a close approach to Antarctica in Tierra Del Fuego (Matallanas et al., 1990).

As a generalization, those species that spawn on gravel beaches tend to occur at higher latitudes than those that spawn on sandy beaches. Estuarine gobies spawn in mud burrows in temperate and tropical waters of the Indo-Pacific, but little is known about the reproduction in many of these species. Beach-spawning fishes in the more northern latitudes tend to migrate in from subtidal habitats. Temperate fishes that spawn on beaches tend to be passive remainers and tide pool emergers, while tropical amphibious fishes include the terrestrially adept skippers (Martin, 1995, 2013). It is likely that additional species and locations will be added to the map as new information is discovered. Opportunities for future comparative studies and field work abound.

Fishes that spawn on beaches are represented throughout the globe and in many different families (see Table 1.1). Biogeographically and phylogenetically diverse, fishes that risk everything to provide the best start in life for their offspring are behaviorally diverse as well. Next, in Chapters 3 and 4, specific reproductive and parental behaviors of beach-spawning fishes will be described.

REFERENCES

Abel, D. C., Koenig, C. C. & Davis, W. P. (1987). Emersion in the mangrove forest fish, *Rivulus marmoratus*: a unique response to hydrogen sulfide. *Environmental Biology of Fishes* 18, 67–72.

ot coddti

me clanly.

Barton, M. G. (1982). Intertidal vertical distribution and diets of five species of central California stichaeoid fishes. *California Fish and Game* 68, 174–182.

Benson, K. R. (2002). The study of vertical zonation on rocky intertidal shores: a historical perspective. *Integrative and Comparative Biology* 42, 776–779.

Breder, C. M. & Rosen, D. E. (1966). *Modes of Reproduction in Fishes*. Garden City: Natural History Press, 941 pp.

Bridges, C. R. (1988). Respiratory adaptations in intertidal fish. *American Zoologist* 28, 79–96.

Bridges, C. R. (1993). Ecophysiology of intertidal fish. In *Fish Ecophysiology* (Rankin, J. C. & Jensen, F. B., eds.), 375–400. London: Chapman & Hall.

Clack, J. A. (2007). Devonian climate change, breathing, and the origin of the tetrapod stem group. *Integrative and Comparative Biology* 47, 510–523.

Cohen, D. M. (1970). How many recent fishes are there? *Proceedings of the California Academy of Sciences* 38, 341–346.

Coleman, R. M. (1999). Parental care in intertidal fishes. In *Intertidal Fishes: Life in Two Worlds* (Horn, M. H., Martin, K. L. M. & Chotkowski, M. A., eds.), 165–180. San Diego: Academic Press.

Courtenay, W. R., Jr., Sahlman, H. F., Miley, W. W. II & Herrema, D. J. (1974). Exotic fishes in fresh and brackish waters of Florida. *Biological Conservation* 6, 292–302.

Crane, J. (1981). Feeding and growth by the sessile larvae of the teleost *Porichthys notatus*. *Copeia* 1981, 895–897.

Davis, J. L. D. (2000). Spatial and seasonal patterns of habitat partitioning in a guild of southern California tidepool fishes. *Marine Ecology Progress Series* 196, 253–268.

Davis, J. L. D. (2001). Diel changes in habitat use by two tidepool fishes. *Copeia* 2001, 835–841.

DeMartini, E. E. (1999). Intertidal spawning. In *Intertidal Fishes: Life in Two Worlds* (Horn, M. H., Martin, K. L. M. & Chotkowski, M. A., eds.), 143–164. San Diego: Academic Press.

Diaz, R. J. & Rosenberg, R. (2008). Spreading dead zones and consequences for marine ecosystems. *Science* 32, 926–929.

Fangue, N. A., Osborne, E. J., Todgham, A. E. & Schulte, P. M. (2011). The onset temperature of the heat-shock protein response and whole-organism thermal tolerance are tightly correlated in both laboratory-acclimated and field-acclimatized tidepool sculpins (*Oligocottus maculosus*). *Physiological and Biochemical Zoology* 84, 341–352. DOI: 10.1086/660113.

Frank, K. T. & Leggett, W. C. (1981). Prediction of egg development and mortality rates in capelin (*Mallotus villosus*) from meteorological, hydrographic, and biological factors. *Canadian Journal of Fisheries and Aquatic Science* 38, 1327–1338.

Gibson, R. N. (1982). Recent studies on the biology of intertidal fishes. *Oceanography and Marine Biology Annual Reviews* 20, 363–414.

Gibson, R. N. (1999). Movement and homing in intertidal fishes. In *Intertidal Fishes: Life in Two Worlds* (Horn, M. H., Martin, K. L. M. & Chotkowski, M. A., eds.), 97–125. San Diego: Academic Press.

Graham, J. B. (1976). Respiratory adaptations of marine air-breathing fishes. In *Respiration in Amphibious Vertebrates* (Hughes, G. M., ed.), 165–187. London: Academic Press.

Graham, J. B. (1997). *Air-Breathing Fishes: Evolution, Diversity and Adaptation*. San Diego: Academic Press.

Graham, J. B. & Lee, H. J. (2004). Breathing air in air: in what ways might extant amphibious fish biology relate to prevailing concepts about early tetrapods, the evolution of vertebrate air-breathing, and the vertebrate land transition? *Physiological and Biochemical Zoology* 77, 720–731.

Graham, J. B., Rosenblatt, R. H. & Gans, C. (1978). Vertebrate air breathing arose in fresh waters and not in the oceans. *Evolution* 32, 459–463.

Griggs, G., Patsch, K. & Savoy, L. (2005). *Living with the Changing California Coast.* Berkeley, CA: University of California Press, 551 pp.

Halpin, P. M. & Martin, K. L. M. (1999). Aerial respiration in the salt marsh fish *Fundulus heteroclitus* (Fundulidae). *Copeia* 1999, 743–748.

Hensley, D. A. & Courtenay, W. R. Jr. (1980). *Clarias batrachus* (Linnaeus) walking catfish. In *Atlas of North American Freshwater Fishes* (Lee, D. S., Gilbert, C. R., Hocutt, C. H., Jenkins, R. E., McAllister, D. E. & Stauffer, J. R. Jr., eds.), 475. North Carolina Biological Survey Publication #1980-12. North Carolina State Museum of Natural History.

Horn, M. H. & Gibson, R. N. (1988). Intertidal fishes. *Scientific American* 256, 64–70.

Horn, M. H. & Martin, K. L. (2006). Rocky intertidal zones. In *Ecology of Marine Fishes: California and Adjacent Waters* (Allen, L. A., Horn, M. H., & Pondella, D., eds.), 205–226. Berkeley, CA: University of California Press.

Horn, M. H., Martin, K. L. M. & Chotkowski, M. A. (eds.). (1999). *Intertidal Fishes: Life in Two Worlds.* San Diego: Academic Press, 339 pp.

Inger, R. F. (1957). Ecological aspects of the origin of tetrapods. *Evolution* 11, 373–376.

Ishimatsu, A. & Gonzales, T. T. (2011). Mudskippers: Front runners in the modern invasion of land. In *The Biology of Gobies* (Patzner, R. A., Van Tassell, J. L., Kovacic, M. & Kapoor, B. G., eds.), 609–638. Enfield, NH: CRC Press, Taylor & Francis.

Ishimatsu, A. & Graham, J. B. (2011). Roles of environmental cues for embryonic incubation and hatching in mudskippers. *Integrative and Comparative Biology* 51, 38–48. DOI: 10.1093/icb/icr018.

Ishimatsu, A., Yoshida, Y., Itoki, N., Takeda, T., Lee, H. J. & Graham, J. B. (2007). Mudskippers brood their eggs in air but submerge them for hatching. *Journal of Experimental Biology* 210, 3946–3954.

Janis, C. & Farmer, C. (1999). Proposed habitats of early tetrapods: gills, kidneys, and the water-land transition. *Zoological Journal of the Linnaean Society* 126, 117–126.

Jones, B. C. (1972). Effect of intertidal exposure on survival and embryonic development of Pacific herring spawn. *Journal of the Fisheries Research Board of Canada* 29, 1119–1124.

Kramer, D. (1978). Terrestrial spawning of *Brycon petrosus* (Pisces: Characidae) in Panama. *Copeia* 1978, 536–537.

Krekorian, C. O. (1976). Field observations in Guyana on the reproductive biology of the spraying characid, *Copeina arnoldi* Regan. *American Midland Naturalist* 96, 88–97.

Leigh, E. G., Jr., Paine, R. T., Quinn, J. F. & Suchanek, T. H. (1987). Wave energy and intertidal productivity. *Proceedings of the National Academy of Science* 84, 1314–1318.

Marliave, J. B. & DeMartini, E. E. (1977). Parental behavior of intertidal fishes of the stichaeid genus *Xiphister*. *Canadian Journal of Zoology* 55, 60–63.

Martin, K. L. M. (1995). Time and tide wait for no fish: intertidal fishes out of water. *Environmental Biology of Fishes* 44, 165–181.

Martin, K. L. M. (1996). An ecological gradient in air-breathing ability among marine cottid fishes. *Physiological Zoology* 69, 1096–1113.

Martin, K. L. M. (1999). Ready and waiting: delayed hatching and extended incubation of anamniotic vertebrate terrestrial eggs. *American Zoologist* 39, 279–288.

Martin, K. L. M. (2013). Review paper: Theme and variations: amphibious air-breathing intertidal fishes. *Journal of Fish Biology* 84, 577–602. DOI:10.1111/jfb.12270.

Martin, K. L. M. & Bridges, C. R. (1999). Respiration in water and air. In *Intertidal Fishes: Life in Two Worlds* (Horn, M. H., Martin, K. L. M. & Chotkowski, M. A., eds.), 54–78. San Diego: Academic Press.

Martin, K. L. & Carter, A. L. (2013). Brave new propagules: terrestrial embryos in anamniotic eggs. *Integrative and Comparative Biology* 53, 233–247. DOI:10.1093/icb/ict018.

Martin, K. L. M. & Strathmann, R. A. (1999). Aquatic organisms, terrestrial eggs: early development at the water's edge. Introduction to the symposium. *American Zoologist* 39, 215–217.

Martin, K. L. M. & Swiderski, D. (2001). Beach spawning in fishes: a phylogenetic approach. *American Zoologist* 41, 526–537.

Martin, K. L. M., Bailey, K., Moravek, C. & Carlson, K. (2011). Taking the plunge: California grunion embryos emerge rapidly with environmentally cued hatching. *Integrative and Comparative Biology* 51, 26–37. DOI: 10.1093/icb/icr037.

Martin, K. L., Moravek, C. L. & Flannery, J. A. (2009). Embryonic staging series for the beach spawning, terrestrially incubating California grunion *Leuresthes tenuis* (Ayres 1860) with comparisons to other Atherinomorpha. *Journal of Fish Biology* 75, 17–38. DOI: 10.111/j.1095–8649.2009.02260.x

Martin, K. L. M., Van Winkle, R. C., Drais, J. E. & Lakisic, H. (2004). Beach spawning fishes, terrestrial eggs, and air breathing. *Physiological and Biochemical Zoology* 77, 750–759.

Martinez, M. L., Landry, C., Boehm, R., Manning, S., Cheek, A. O. & Rees, B. B. (2006). Effects of long-term hypoxia on enzymes of carbohydrate metabolism in the Gulf killifish, *Fundulus grandis*. *Journal of Experimental Biology* 209, 3851–3861.

Matallanas, J., Rucabado, J., Lloris, D. & Pilar Olivar, M. (1990). Early stages of development and reproductive biology of South American eelpout *Austrolycus depressiceps* Regan, 1913 (Teleostei: Zoarcidae), *Sciencias Marinas* 54, 257–261.

McLachlan, A. & Brown, A. C. (2006). *The Ecology of Sandy Shores,* 2nd Edition. San Diego, CA: Academic Press, 373 pp.

Meller, C. L., Meller, R., Simons, R. P. & Podrabsky, J. E. (2014). Patterns of ubiquitylation and SUMOylation associated with exposure to anoxia in embryos of the annual killifish. *Journal of Comparative Physiology B* 184, 235–247.

Middaugh, D. (1981). Reproductive ecology and spawning periodicity of the Atlantic silverside, *Menidia menidia* (Pisces: Atherinidae). *Copeia* 1981, 766–776.

Middaugh, D. P., Kohl, H. W. & Burnett, L. E. (1983). Concurrent measurement of intertidal environmental variables and embryo survival for the California grunion, *Leuresthes tenuis*, and Atlantic silverside, *Menidia menidia* (Pisces: Atherinidae). *California Fish & Game* 69, 89–96.

Moring, J. R. (1986). Seasonal presence of tidepool fishes in a rocky intertidal zone of northern California, USA. *Hydrobiologia* 134, 21–27.

Nakamura, R. (1976). Temperature and the vertical distribution of two tidepool fishes (*Oligocottus maculosus, O. snyderi*). *Copeia* 1976, 143–152.

Nakano, K. & Iwama, G. K. (2002). The 70-kDa heat shock protein response in two intertidal sculpins, *Oligocottus maculosus* and *O. snyderi*: relationship of hsp 70 and thermal tolerance. *Comparative Biochemistry and Physiology A* 133, 79–94.

Niedzwiedzki, G., Szrek, P., Narkiewicz, K., Narkiewicz, M. & Ahlberg, P. E. (2010). Tetrapod trackways from the early Middle Devonian period of Poland. *Nature* 463, 43–48.

Penttila, D. E. (2001). Intertidal spawning ecology of three species of marine forage fishes in Washington State. *Journal of Shellfish Research* 20, 1198.

Pfister, C. A. (1992). Sculpin diversity in tidepools. *Northwest Environment* 8, 156–157.

Podrabsky, J. E., Carpenter, J. F. & Hand, S. C. (2001). Survival of water stress in annual fish embryos: dehydration avoidance and egg envelope amyloid fibers. *American Journal of Physiology* 280, R123–R131.

Podrabsky, J. E., Lopez, J. P., Fan, T. W. M., Higashi, R. & Somero, G. N. (2007). Extreme anoxia tolerance in embryos of the annual killifish *Austrofundulus limnaeus*: insights from a metabolomics analysis. *Journal of Experimental Biology* 210, 2253–2266. DOI:10.1242/jeb.005116.

Podrabsky, J. E., Tingaud-Sequeira, A. & Cerda, J. (2010). Metabolic dormancy and responses to environmental desiccation in fish embryos. In *Dormancy and Resistance in Harsh Environments* (Lubzens, E., Cerda, J. & Clark, M., eds.), 203–225. Berlin: Springer-Verlag. DOI: 10.1007/978-3-642-12422-8_12.

Randall, D. J., Burggren, W. W., Farrell, A. P. & Haswell, M. S. (1981). *The Evolution of Air-Breathing in Vertebrates.* London: Cambridge University Press, 133 pp.

Rice, C. A. (2006). Effects of shoreline modification on a northern Puget Sound beach: microclimate and embryo mortality in surf smelt (*Hypomesus pretiosus*). *Estuaries and Coasts* 29, 63–71.

Richkus, W. A. (1978). A quantitative study of intertidepool movement of the wooly sculpin, *Clinocottus analis. Marine Biology* 49, 227–284.

Romer, A. S. (1967). Major steps in vertebrate evolution. *Science* 158, 1629–1637.

Sayer, M. D. J. & Davenport, J. (1991). Amphibious fish: why do they leave water? *Reviews in Fish Biology and Fisheries* 1, 159–181.

Schultze, H. P. (1999). The fossil record of the intertidal zone. In *Intertidal Fishes: Life in Two Worlds* (Horn, M. H., Martin, K. L. M. & Chotkowski, M. A., eds.), 373–392. San Diego: Academic Press.

Shimizu, N., Sakai, Y., Hashimoto, H., & Gushima, K. (2006). Terrestrial reproduction by the air-breathing fish Andamia tetradactyla (Pisces: Blenniidae) on supralittoral reefs. *Journal of Zoology* 269, 357–364.

Sloman, K. A., Mandic, M., Todgham, A. E., Fangue, N. A., Subrt, P. & Richards, J. G. (2008). The response of the tidepool sculpin, *Oligocottus maculosus*, to hypoxia in laboratory, mesocosm and field environments. *Comparative Biochemistry and Physiology* 149A, 284–292.

Stevenson, T. A. & Stevenson, A. (1972). *Life between Tidemarks on Rocky Shores.* San Francisco: W. H. Freeman & Co, 425 pp.

Strathmann, R. R. & Hess, H. C. (1999). Two designs of marine egg masses and their divergent consequences for oxygen supply and desiccation in air. *American Zoologist* 39, 253–260.

Summers, K., McKeon, C. S. & Heying, H. (2006). The evolution of parental care and egg size: a comparative analysis in frogs. *Proceedings of the Royal Society B* 273. DOI: 10.1098/rsbp.2005.3368.

Swenson, R. O. (1997). Sex-role reversal in the tidewater goby, *Eucyclogobius newberryi. Environmental Biology of Fishes* 50, 27–40. DOI:10.1023/A:1007352704614.

Taylor, D. S. (2012). Twenty-four years in the mud: what have we learned about the natural history and ecology of the mangrove rivulus, *Kryptolebias marmoratus*? *Integrative and Comparative Biology* 52, 724–736.

Taylor, D. S., Turner, B. J., Davis, W. P. & Chapman, B. B. (2008). A novel terrestrial fish habitat inside emergent logs. *American Naturalist* 171, 263–266.

Taylor, M. H. (1999). A suite of adaptations for intertidal spawning. *American Zoologist* 39, 313–320.

Taylor, M. H., DiMichele, L. & Leach, G. J. (1977). Egg stranding in the life cycle of the mummichog, *Fundulus heteroclitus. Copeia* 1977, 397–399.

Tewksbury, H. T. & Conover, D. O. (1987). Adaptive significance of intertidal egg deposition in the Atlantic silverside *Menidia menidia. Copeia* 1987, 76–83.

Thibaudeau, G. & Altig, R. (1999). Endotrophic anurans: development and evolution. In *Tadpoles: The Biology of Anuran Larvae* (McDiarmid, R. W. & Altig, R., eds.), 170–188. Chicago: University of Chicago Press.

Thomson, D. A. & Muench, K. A. (1976). Influence of tides and waves on the spawning behavior of the Gulf of California grunion, *Leuresthes sardina* (Jenkins and Evermann). *Bulletin of the Southern California Academy of Sciences* 75, 198–203.

Truchot, J. P. & Duhamel-Jouve, A. (1980). Oxygen and carbon dioxide in the marine intertidal environment: diurnal and tidal changes in rockpools. *Respiration Physiology* 39, 241–254.

Walker, B. (1952). A guide to the grunion. *California Fish and Game* 38, 409–420.

Wells, A. W. (1986). Aspects of ecology and life history of the woolly sculpin, *Clinocottus analis*, from Southern California. *California Fish and Game* 72, 213–226.

Williams, G. C. (1954). Differential vertical distribution of the sexes in *Gibbonsia elegans* with remarks on two nominal subspecies of this fish. *Copeia* 1954, 267–273.

Williams, G. C. (1957). Homing behavior of California rocky shore fishes. *University of California Publications in Zoology* 59, 249–284.

Yamahira, K. (1996). The role of intertidal egg deposition on survival of the puffer, *Takifugu niphobles* (Tetraodontidae). *Environmental Biology of Fishes* 40, 255–261.

Yoshiyama, R. M., Gaylord, K. B., Philippart, M. T., Moore, T. R., Jordan, J. R., Coon, C. C., Schalk, L. L., Valpey, C. J. & Tosques, I. (1992). Homing behaviour and site fidelity in intertidal sculpins (Pisces: Cottidae). *Journal of Experimental Marine Biology and Ecology* 160, 115–130.

Yoshiyama, R. M., Sassaman, C. & Lea, R. N. (1986). Rocky intertidal fish communities of California: temporal and spatial variation. *Environmental Biology of Fishes* 17, 23–40.

Yoshiyama, R. M., Valey, C. J., Schalk, L. L., Oswald, N. M., Vaness, K. K., Lauritzen, D. & Limm, M. (1995). Differential propensities for aerial emergence in intertidal sculpins (Teleostei; Cottidae). *Journal of Experimental Marine Biology and Ecology* 191, 195–207.

Zander, C. D., Nieder, J. & Martin, K. L. M. (1999). Vertical distribution patterns. In *Intertidal Fishes: Life in Two Worlds* (Horn, M. H., Martin, K. L. M. & Chotkowski, M. A., eds.), 26–53. San Diego: Academic Press.

3 Locals Only: Beach-Spawning Behavior in Resident Intertidal Fishes

Producing the next generation is one of the great challenges of life. Each of the astonishing variety of species on Earth can be distinguished in part by their reproductive isolation from other species. The diversity of behaviors and morphological adaptations of fishes for reproduction are vast and endlessly fascinating, particularly in the case of these fishes that choose to spawn on beaches, in an environment that is peripheral to, or in fact different from, the environment in which they spend most of their larval and adult lives. This chapter and the next will describe the reproductive adaptations and the rich tapestry of behaviors of beach-spawning fishes, as revealed in the many diverse species that reproduce on beaches.

Resident intertidal fishes enjoy beachfront property and are found within the intertidal zone year-round (Gibson, 1982; Horn et al., 1999). Fishes that reside within the intertidal zone show a variety of forms of courtship, mating, and parental care. These fishes live in the same location throughout the year and simply alter their behavior during reproductive efforts. In this chapter, specific examples of beach-spawning behavior will be described for species from many different lineages, for rocky, muddy, or estuarine intertidal habitats.

3.1 MANY TELEOSTS REPRODUCE WITH PELAGIC EGGS

Compared with mammals, teleost fishes in general have relatively basic reproductive organs. Ovaries and testes produce eggs and sperm, which are delivered through ducts to the outside. Accessory glands may produce substances that enhance sperm motility, increase adhesion of eggs to substrates, chemically attract mates and gametes to one another, or provide antimicrobial protection to the eggs (Giacomello et al., 2006). Either gender may have a genital papilla that can guide the release of gametes during spawning, allowing females to place individual eggs on a substrate and males to control ejaculation and fertilize multiple clutches (Aryafar, 2012). Eggs and sperm typically are viable for only a few minutes outside of the body before fertilization, so it is important that the gametes combine quickly and efficiently.

In the open ocean, teleost fishes typically broadcast their pelagic eggs and sperm, releasing their gametes into the water column to interact and then float freely in the plankton after external fertilization takes place at the whim of the current. Fishes may engage in spawning rushes, releasing clouds of sperm that must locate their egg targets rapidly before they disappear. Fishes that broadcast pelagic eggs may spawn in groups and have little choice about what happens next to their offspring, including which father provides the other half of the embryos' genetic material. Female fishes with demersal, adherent eggs may have more control over the fate of these propagules, not only in protecting the developing embryos with a safe oviposition site but also in terms of her mate. A clutch of eggs placed in a stable location can be defended from other, less desirable mates and hidden from some potential predators.

Teleost fishes that are pelagic spawners typically produce numerous small eggs, from hundreds to hundreds of thousands in a clutch. The pelagic eggs of these species float in the water column after fertilization while they undergo rapid development. The embryos hatch within a few days into tiny planktonic larvae. Rapid development to hatching is critical, as these nonmotile embryos are defenseless against numerous predators. With the small size of each individual egg, the mother is able to produce large numbers (Blaxter, 1988), so even though high mortality of eggs and embryos occurs, the odds are that some individuals will survive. The larvae usually hatch at an early stage of development and must begin feeding soon afterward, because the energy stored in what remains of the small yolk is used up quickly.

3.2   BEACH-SPAWNING FISHES PRODUCE DEMERSAL EGGS

The demersal eggs of beach-spawning fishes are generally larger in diameter than pelagic eggs, with greater energy content. Demersal eggs do not float but tend to sink to the bottom in water and stay in one place on the substrate. For more information about demersal eggs, see Chapter 7.

Fishes with demersal eggs may spawn at a more leisurely pace. Females may carefully place their eggs either individually or in clusters such as a self-adherent sphere or a monolayer attached to some structure, perhaps a rock, mussel shell, or some vegetation (Figure 3.1). During this process, she may be closely followed by one or more attendant males that release sperm to fertilize the eggs. In species with copulation and either internal fertilization or internal gamete association, males may be absent when oviposition occurs. Demersal eggs are considered "sticky" because of tiny filaments on the chorion in many species that allow attachment to other eggs or to a substrate (Rizzo et al., 2002; see Chapter 7).

One advantage of beach spawning for marine species is tidal air exposure of the developing eggs. Therefore, initial oviposition and stability in the intended location are vital to their survival (Tewksbury & Conover, 1987; Yamahira, 1996; Martin et al., 2004). Some delay hatching and extend incubation, waiting to hatch in response to an environmental cue (Martin, 1999; Martin et al., 2011).

Fishes that spawn on beaches may be restricted to spawning during the highest high tides at the full and new moon in order to obtain access to preferred spawning substrates (Taylor, 1984). The access to a substrate is more dependent on the tides for some species than for others. For most beach-spawning fishes, spawning is aquatic, although there are exceptions. Spawning often takes place during the slack water period as the tide turns from flood to ebb. This provides the necessary calm conditions for external fertilization and oviposition. See Chapter 2 for more on tidal cycles.

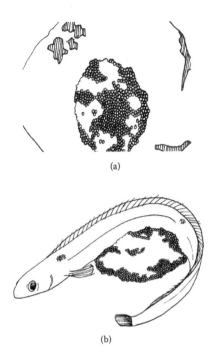

FIGURE 3.1   Beach-spawning fishes produce demersal eggs that may be (a) placed on a biogenic substrate such as a mussel shell or Spartina, or placed directly onto a rocky substrate to which they adhere; or (b) adhere to one another, but not the substrate, in a spherical clutch.

3.3   DIVERSE MATING BEHAVIORS AND MATING SYSTEMS ARE SEEN IN BEACH-SPAWNING FISHES THAT ARE INTERTIDAL RESIDENTS

Intertidal spawning has evolved in a diversity of lineages. A variety of mating systems are seen in beach-spawning fishes that reside in the intertidal zone. These range from monogamy to polyandry, to harems and sneaker males. Some beach-spawning fishes form pair bonds that last only as long as the spawning act; others mate and then interact throughout the incubation of the eggs. In some cases the strongest bond may be between a parent and a clutch of eggs in a nest, rather than between two parents. Table 3.1 summarizes the types of mating systems found among beach-spawning fishes, with examples. Additional details follow below.

Males of some species defend multiple clutches within one nest, each produced by a different female. If the male is guarding a desirable nest site, he is even more appealing to females when he is already guarding developing eggs (Petersen, 1989). In some species, females hedge their bets by depositing small numbers of eggs into multiple locations when spawning. Each partial clutch may be fertilized by the same male or different males. Using eggs as courtship devices is seen in many species of fishes, and this may be effective for a male whether or not he adopted clutches fertilized by other males, and, somewhat alarmingly, even if some eggs in his care

TABLE 3.1

Mating Systems Seen in Beach-Spawning Resident Intertidal Fishes

Type, Brief Description	Example Beach-Spawning Fish Species
Monogamy, biparental care of eggs	*Pholis laeta, Typhlogobius californiensis*
Copulation, internal gamete association (IGA)	*Oligocottus maculosus, Clinocottus analis*
With male egg guarding	*Artedius lateralis, Artedius harringtoni*
Nest parasitism on another species	*Asemichthys taylori* on *Enophrys bison* eggs
Polygyny, territorial males guard multiple clutches	*Gobiesox meandricus, Andamia tetradactylus*
With "sneaker" males	*Axoclinus nigricaudatus*
Monogamy, sequential hermaphrodites	*Lythrypnus dalli*
Group spawning, multiple males and multiple females	*Aschelichthys rhodorus*
Internal fertilization with viviparity	*Anableps anableps*

Note: See texts of Chapters 3 and 4 for additional details on each species.

are cannibalized (Petersen et al., 2005). The filial cannibalism may help protect the clutch if it removes dead or diseased progeny, or if it allows a guarding male to maintain his energy without leaving the nest to hunt for food.

For some species, eggs attached to a boulder or other substrate may attract other nonterritorial males as well as females. Particularly for those species with territorial males holding nest sites, unequal success in mating occurs between males within the population. A few males with desirable nest sites may be able to mate with multiple females. An alternative reproductive tactics (ART) for males is to avoid direct competition with larger territorial males when small and invest more energy into producing large testes. These small males are more like females in appearance.

The alternative males can sneak into a nest site guarded by a larger male, masquerading as a female. When the territorial male induces a female to spawn in his nest, the "sneaker" male releases milt over the eggs at the same time, producing some offspring of his own. This strategy is seen in an intertidal resident, the triplefin blenny *Axoclinus nigricaudus* (Neat, 2001; Hastings & Petersen, 2010), and a beach-spawning migrant, the plainfin midshipman *Porichthys notatus* (Brantley & Bass, 1994). See Chapter 4 for more on migrant beach-spawning fishes.

Whether or not there is a lasting pair bond, parental care for a clutch may be provided by either male or the female fish. Both partners may care for the incubating embryos in some species (Coleman, 1999). Of course, as is the situation for most fishes that spawn in the open ocean pelagic waters, many fishes that spawn on beaches have no additional parental involvement with their young once fertilization occurs.

Among the many species of teleost fishes are some species that are sequential hermaphrodites, with individuals that can change from female to male (protogyny), or male to female (protandry), depending on the circumstances of their environment (Warner, 1988). In addition, simultaneous hermaphroditism is seen in some fish species that form pair bonds. Each partner provides sperm for the other's eggs, switching roles over a period of time (Munday et al., 2010). Neither sequential nor simultaneous hermaphroditism has been described to date for beach-spawning fishes, but one potential candidate is the blueband goby *Lythrypnus dalli* (Rodgers et al., 2007). This species spawns in the high subtidal zone and occasionally in the low intertidal zone. See Chapter 4 for more details.

Internal fertilization is rare among teleost fishes, but copulation occurs in some resident intertidal fishes. Although one might presume that copulation precedes internal fertilization, this is not necessarily the case. In many cottid species, males deposit sperm within the female's reproductive tract during copulation, but fertilization may not happen until the eggs leave her body during oviposition (Munehara et al., 1989). Because of this internal gamete association (IGA), sperm may be stored briefly before fertilization, with one or more sperm in association with the micropyle of each egg (Petersen et al., 2005). When the female lays her eggs, the sperm are carried along and fertilization occurs. Some eggs may be fertilized externally as well, and this may be by the same or different males. The males swim over the eggs on the nest and provide additional milt. IGA without fertilization delays the initiation of embryonic development until oviposition occurs (Munehara et al., 1989, 1991).

3.4 FISHES THAT RESIDE IN THE ROCKY INTERTIDAL ZONE SPAWN ON THE ROCKY BEACH

Resident fishes that live and spawn in the rocky intertidal zone include many species of cottids, stichaeids, liparids, blenniids, tripterygiids, pholids, and zoarcids. These are typically small, cryptic fishes that generally live in rock pools or under boulders. Many tide pool fishes can emerge from water for terrestrial activity and breathe air (Martin, 1993; Martin & Bridges, 1999). In the intertidal zone, emergence by amphibious fishes typically occurs during low tides, not during spawning. Intertidal spawning behavior typically occurs around the high tides, when aquatic conditions prevail (DeMartini, 1999). See Chapter 5 for more about amphibious emergence behavior and air breathing in beach-spawning fishes.

Many sculpins (Cottidae) have intromittent organs for copulation. These are so different that they are diagnostic for each species. In general cottid females lay brightly colored eggs attached to substrates. Some cottid species produce eggs with toxins that, like the bright colors, are derived from the female's diet. Most intertidal cottid species show little or no parental care.

Intertidal sculpins spawning in the rocky intertidal zone appear to have two general forms of mating systems: either IGA with no parental care, or external fertilization with some incidental male parental care (Muñoz, 2010). The bonds formed during copulation are temporary, and individuals in most cottid species spawn with multiple partners. A male may use the incubating eggs already present in his nest to attract multiple females. She may perceive these as evidence of his fitness, or perhaps the presence of multiple clutches appears to dilute the risk of predation risk for each (Petersen et al., 2005; Petersen, 1989). Sculpin nests are laid directly on the ceiling of their nesting sites, the underside of boulders, sometimes with multiple clutches on top of one another (Andrew Kinziger, personal communication).

The tide pool sculpin *Oligocottus maculosus* copulates with an intromittent organ and deposits sperm into the female's reproductive tract for IGA (see Section 3.3), and fertilization takes place after oviposition rather than internally (Petersen et al., 2005). *O. maculosus,* the most common sculpin seen in tide pools along the west coast of North America, may spawn twice a year, producing bright green or red eggs depending on the type of algae consumed (Pierce & Pierson, 1990). After copulation, 100 to 700 eggs are produced, but no parental care occurs.

Other intertidal sculpins copulate and, like *O. maculosus,* show IGA and external fertilization. While copulation requires at least temporary mate choice, Morris (1956, p. 314) described spawning of the fluffy sculpin *Oligocottus snyderi* "in an atmosphere of carefree promiscuity." The woolly sculpin *Clinocottus analis* female can store sperm after copulation before producing eggs that develop without any additional male contact for up to two months (Hubbs, 1966).

The rosylip sculpin, *Ascelichthys rhodorus,* has a reproductive strategy unlike any other sculpin (Petersen et al., 2004). Spawning aggregations form near boulders in the low intertidal zone, and large groups of males congregate near spawning sites. External fertilization occurs during group spawning as females deposit their clutches, indicating sperm competition is likely. The males neither guard individual nests nor perform parental care, and may leave before the eggs hatch. Nevertheless, the male appears to provide some incidental protection to the developing clutches, simply by his temporary presence (Petersen et al., 2004).

Three sympatric cottid species of *Artedius* found in the intertidal zone of the northeast Pacific can be distinguished by differences in their reproductive behavior. For all three *Artedius* species, larger males tend to be more successful in attracting females. These males guard their eggs (Ragland & Fischer, 1987; Petersen et al., 2005), an unusual situation for fishes with either internal fertilization or IGA (Clutton-Brock, 1991). Males are larger and more brightly colored than females, and more numerous at spawning sites. In addition, males guarding eggs enjoy an advantage in obtaining additional matings (Petersen et al., 2005; Petersen & Hess, 2011).

All three species, *Artedius lateralis, A. fenestralis,* and *A. harringtoni,* are capable of both IGA and external fertilization to initiate development. *A. harringtoni* is more likely to employ copulation with IGA than the other two species are. Perhaps this is related to the fact that the male intromittent organ of *A. harringtoni* is 10 times longer than that of its two congeners (Petersen et al., 2005). For *A. harringtoni* the sperm are motile at osmolarities lower than seawater in osmolarities more similar to conditions within the female's body. Other cottid species that rely on IGA also have greater sperm motility at lower osmolarities (Munehara et al., 1997). In contrast, the higher osmolarity of seawater conditions enhances the motility of sperm for species that have external fertilization.

One cottid species, *Hemilepidotus gilberti,* produces dimorphic sperm; the non-functional paraspermatozoa help to guide the functional euspermatozoa toward the eggs and may also help to protect paternity by limiting sperm competition between males (Hayakawa et al., 2002, 2004). This species is found in the intertidal zone, but it is not clear whether it spawns intertidally.

Closely related to the Cottidae, the snailfish family Liparidae also contains species that live from the intertidal zone to deep in the water. These small fishes nest in biogenic substrates such as bivalve shells, barnacle colonies, and worm tubes, as well as hydroids and algae. One species even lays eggs within the gill cavity of a crab (Munoz, 2010). However, little is known about the spawning patterns of resident intertidal liparid species.

Blennids (Blenniidae) are a very successful speciose group worldwide. Male blennies attract females with visual cues, chemical cues, and sounds, and typically guard oviposition sites and provide parental care of eggs (Barata et al., 2008; Hastings & Petersen, 2010; Petersen & Hess, 2011). Intertidal species are relatively small, to 12 cm long. Both intertidal and subtidal blennid species are territorial and guard cavity nests (Stephens et al., 1970). The males choose sites among rocks, barnacle shells, or other objects. Males then engage in courtship displays to attract females by swimming up and down near the nest entrance with fins spread, and by nodding the head to lead the female to the nest entrance (Almada & Santos, 1995). A female enters the nest to spawn, followed by the male. Male parental care in the form of nest guarding is typical for blennies (Almada & Santos, 1995; Coleman, 1999). The care consists of aeration of the eggs by fanning them with the fins, keeping the egg surfaces clean by swimming over them and removing debris, and defending against predators.

The intertidal blennids *Lipophrys pholis*, *Coryphoblennius galerita*, and *Salarias pavo* spend more time with the nests and less time feeding than subtidal species do (Goncalves & Almada, 1998). During courtship display, subtidal species show more vertical movement than intertidal species can manage in the shallow water of a tide pool and greater turbulence from waves at high tide (Almada & Santos, 1995). Intertidal nests are submerged by tides daily, yet the nests of these species are usually emerged into air for more hours than they are submerged.

The zebra blenny *Istiblennius zebra* of Hawaii nest in cavities or crevices formed by natural holes in the lava rock. The male further excavates the space by removing sand or other material with his mouth and dropping it outside the nest (Phillips, 1977). The tidal excursion around the Hawaiian Islands is relatively small; nevertheless the shallow lava rock pools are separated from the open ocean and refilled only by frequent waves. The nest site is supralittoral and above the reach of all but the highest tides, although the eggs and adults are rarely exposed to air.

During aquatic courtship, the *I. zebra* male swims in tight vertical circles to attract the attention of a female, nodding his head toward the opening of the nest as if to invite her inside (Phillips, 1977). If she seems interested, he leads her toward the nest with a behavior called "dip-swim," in which he positions himself vertically for a brief period, then dips down to a horizontal position and swims a few centimeters, then raises himself up vertically once more. At the nest, the male swims in circles around the entrance, and both partners enter the nest to mate. The female turns so her head is emerged from the nest, and begins laying eggs along the sides of the cavity. The male quivers beside her, releasing his milt. Oviposition may take many minutes, during which the eggs produced may completely cover the inside walls of the enclosure. After the female leaves, the male guards the eggs by fanning them during their incubation (Phillips, 1977).

The rockskipper blenny *Alticus kirki* lives most of its life above the water line (Brown et al., 1992). It nests in cavities above the water line, displaying dramatic courtship behavior out of water and territorial defense of the nest site, although little has been published about it (Zander et al., 1999). Adult *A. kirki* can cling to vertical rock faces because of special hooklets on the fins (Zander, 1972), and can maintain its position in the littoral zone even in the presence of pounding waves. A congener, the leaping blenny *Alticus saliens* in the Indo-Pacific, is also very terrestrial. It nests above the water line (Abel, 1973), but little is known about its reproduction beyond that the eggs are demersal and adhesive (Breder & Rosen, 1966).

Spawning for the rockhopper blenny *Andamia tetradactyla* is exceptional as it takes place out of water, at both rising and falling tides, when waves splash the nest site but while it is emerged above the water line (Shimizu et al., 2006). This may be one of the most amphibious examples of beach spawning among rocky intertidal fishes. *A. tetradactyla* is found in the Ryuku Islands of Japan, where it inhabits the high intertidal zone and feeds on algal mats. On rising tides, these small fish move upward above the water line, "as if escaping from the flooding water" (Shimizu et al., 2006, p. 359). This same behavior also occurs in *Alticus kirki* (Brown et al., 1992). Nesting occurs in crevices between rocks or artificial structures at the high end of the intertidal zone, so that nests are emerged from water up to 12 h a day, only submerged by the highest tides. Courtship, as with other blennies, involves a territorial male nodding toward a female and leading her by indicating the entrance of the nest (Figure 3.2a). All the while, he changes color of the head and body from dark to pink, red, and yellow. Mating females become white or yellow.

Captive *A. tetradactyla* in aquariums have been observed mating (Shimizu et al., 2006). A nest chamber was made available under humid conditions but out of water. The female oviposited on either the walls or ceiling of the nest chamber, whichever was broader, clinging to the moist wall with suction of the fins and body. The eggs were placed in a monolayer, extruded along with a whitish substance that seems to be adhesive (Figure 3.2b). Meanwhile, the male guarded the nest entrance during oviposition (Figure 3.2b). After the female is finished, he pressed his body against the egg mass to rub his genital papilla from side to side over the area, providing milt (Shimizu et al., 2006). This form of fertilization by contact with eggs a few minutes after oviposition is also seen in other blennies (Giacomello et al., 2008).

Eggs of *A. tetradactyla* in the natural habitat are submerged tidally on a daily basis, but are able to develop fully out of water in an aquarium if kept humid. Embryos are ready to hatch within about 7 to 10 days but do not hatch until the eggs are submerged in water, for example as during a high tide. See Chapter 8 for more on hatching. Each male guards his nest and may mate with multiple females, so that the nest may contain embryos from clutches at several different stages of development. The enclosure of the nest cavity retains more humidity than the surrounding air. The male seems to provide extra protection to the clutch by remaining with it throughout incubation, occasionally placing his body gently over the eggs during times of air emergence (Shimizu et al., 2006).

Moving on to a related family of Blennioid rocky intertidal residents, males of the triplefin blenny *Axoclinus nigricaudus* (Tripterygiidae) hold territories on rocky substrate in the Gulf of California, over which they spawn with multiple partners (Neat, 2001; Hastings & Petersen, 2010). *A. nigricaudus* spawns in the shallow subtidal, and eggs are occasionally emerged during low tides. Larger males may hold

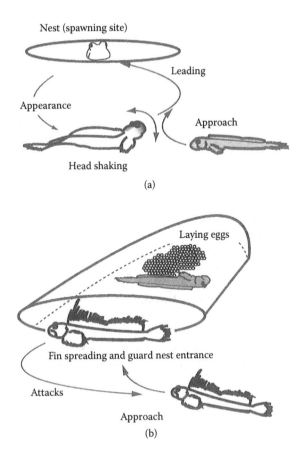

Nest (spawning site)

Leading

Appearance

Approach

Head shaking

(a)

Laying eggs

Fin spreading and guard nest entrance

Attacks

Approach

(b)

FIGURE 3.2 Terrestrial oviposition and fertilization of blenny *Andamia tetradactyla*. (a) The male invites the female into the nest; (b) the male guards the nest while the female oviposits. (From Shimizu, N. et al., *Journal of Zoology*, 269, 2006, 357–364. Used with permission.)

their own nesting sites, but smaller males use alternative reproductive tactics (ART). A smaller male may sneak into a nesting site during an encounter between the dominant male and a female to release his sperm as her eggs are laid (Neat, 2001; Hastings & Petersen, 2010). The smaller males differ in appearance from the territorial males. They have minimal secondary sexual characteristics but greatly enlarged testes.

A. nigricaudatus nests out on an open substrate over short, tufted algal mats (Hastings & Petersen, 2010). The eggs are loosely attached within the algae. This open site is difficult to defend from the sneaker males. Sneaking ART strategy is unusual for species that nest in cavities, as these locations are more defensible by the resident males.

The European intertidal triplefin *Tripterygion notatus* shows similar reproductive behavior, as territorial males guard territories and mate with multiple females (Wirtz, 1978). Eggs already in the nest attract other females (Petersen, 1989). In another triplefin species, *Axoclinus storeyae*, the presence of spawning females attracts other females (Petersen, 1989).

Clingfishes (Gobiesocidae) are sedentary fishes that attach to rocks with a large sucker, modified from the pelvic fins and skinfolds. Intertidal clingfishes attach adhesive eggs in a monolayer to a sheltering rock or the underside of a boulder (Coleman, 1999; Pires & Gibran, 2011). The lappetlip clingfish *Gobiesox barbatulus* of Brazil shows elaborate courtship behavior when two individuals move close together in a nesting area (Pires & Gibran, 2011). The pair may line up nearly touching, side-by-side facing the same direction or opposite directions, or one may line up diagonally across the other, prior to the initiation of mating. Oviposition is indicated by the pair slowly moving close together over a bare area of the substrate.

After oviposition and external fertilization, one parent, presumably the male, initiates parental care behaviors including nest guarding, fanning, and mouthing of the eggs to clean them (Pires & Gibran, 2011). Repeated spawnings may occur between the same pair in a single day over substrate that does not already contain eggs, and nests may contain multiple clutches with embryos at different stages of development.

Male northern clingfish *Gobiesox maeandricus* can be found under rocks in the intertidal zone of the western United States and Canada, with eggs in a monolayer attached to the ceiling of the small, flat chamber formed beneath a boulder (Marliave & DeMartini, 1977; Coleman, 1999). Parental care has not been described; however, these fathers remain with their clutch even when disturbed by humans turning over their boulder homes.

Stichaeid fishes (Stichaeidae) are elongate perciform fishes found in the low to high rocky intertidal zone (Eschmeyer et al., 1983). Their spawning sites are shallow hollows beneath boulders that may be dewatered and emerged into air during low tides. The depression beneath the boulder retains a shallow puddle during low tides, insufficient to cover the adult fish but presumably helpful to maintain humidity within the chamber (Horn & Riegle, 1981). Parental care in stichaeids is common and may be done by the female, the male, or both, evidenced by the adult coiling about a clutch formed by a ball of adhesive eggs (Miki et al., 1987; Coleman, 1992, 1999). The female of the cockscomb prickleback *Anoplarchus purpurescens* guards her clutch of eggs; these adhere to one another in a ball but are not attached to the substrate (Coleman, 1992; Strathmann & Hess, 1999). (See Figure 3.1b.)

Not much is known about spawning activities in stichaeids except the time of year. Individual females may have only one clutch per year (Marliave, 1977). Male *Xiphister atropurpureus* and *X. mucosus* care only for eggs they fertilized, but may tend up to three clutches in the nest, indicating that the males spawn repeatedly with different females (Marliave, 1977).

Pholids (Pholididae) are elongate perciform fishes similar in appearance and closely related to stichaeids. These small intertidal fishes show biparental care by wrapping around egg masses (Qasim, 1957; Coleman, 1999). The sharing of parental duties indicates a robust pair bond, although little is known of the spawning behavior for this group. Two species, *Pholis laeta* and *P. ornata*, live sympatrically near surfgrass beds in British Columbia, Canada. Both are found in the intertidal and subtidal zones in summer months. *P. laeta* breeds intertidally and *P. ornata* breeds subtidally in locations where the two species co-occur, but in areas where *P. laeta* is absent, *P. ornata* breeds intertidally (Hughes, 1985). This observation suggests interspecific competition for nest sites and supports the hypothesis that intertidal oviposition is advantageous for embryos. However there may be some physical feature that both excludes breeding *P. laeta* and alters the oviposition behavior of *P. ornata* at sites where only *P. ornata* occurs.

Another group of slender perciform fishes is the eelpouts (Zoarcidae). Different species show parental care by one or both parents in the intertidal zone. In Argentina's Tierra Del Fuego in the Beagle Channel, a female eelpout *Austrolycus depressiceps* guarding a clutch of 465 eggs was collected from a small cavity in the rocks of the intertidal zone (Matallanas et al., 1990). The demersal eggs, 9.2–9.8 mm in diameter, are the largest marine teleost eggs known. These eggs were completely covered in Cyanophyceae, blue-green algae, and heavily infested with leeches (Matallanas et al., 1990).

Another intertidal zoarcid, *Phucocoetes latitans,* like most tide pool fishes is relatively small, 82–108 mm. It still produces very large eggs, 4.5 mm in diameter, albeit only one or two dozen (Gosztonyi, 1977). Other South American intertidal zoarcids include *Dadyanos insignis* and *Iluocoetes elongatus* (Gosztonyi, 1977). On the opposite side of the world, in the Sea of Ohkotsk, *Magadania kopetski* eelpouts are found intertidally under rocks during the breeding season (Shinohara et al., 2004).

Most zoarcids are deepwater fishes, and little is known about their lives or their reproduction. In the laboratory, captive deepwater zoarcids were observed burrowing and engaging in nesting behavior on substrates. This suggests the possibility that some zoarcids may undertake spawning migrations to nesting areas that contain the appropriate sediment or substrate (Ferry-Graham et al., 2007). More study is needed to find where these species may spawn in their natural habitats.

3.5 FISHES IN ESTUARIES SPAWN ON BEACHES AT THE WATER'S EDGE OR ON INTERTIDAL MUD FLATS

Intertidal habitats in sheltered locations such as bays develop depositional substrates of finer sediments, muds, and clays, rather than the erosional coarse sand, gravel, and rocky areas found on open coastlines. Estuaries are highly productive areas where freshwater meets the ocean, with low wave action (Little, 2000). The muddy sediments may support large areas of seagrasses in the temperate zone or mangrove forests in subtidal and tropical regions. These biogenic substrates provide habitat and food for any species that can tolerate the rapidly changing aquatic conditions.

Estuaries are nutrient rich but highly variable in salinity, temperature, water depth, aquatic oxygen content, and current strength and direction of flow. These effects are caused by tidal cycles, combined with freshwater outflow from rivers and creeks and the shape and depth of the ocean opening (Little, 2000). Adding to the mutability of the environment, plants and animals in the upper intertidal zone are regularly emerged into air by the tides. See Chapter 2 for more on the physical environment.

Clearly, beach-spawning fishes and embryos in estuaries are highly specialized to tolerate a wide spectrum of challenging habitat conditions. On the Atlantic coast of North America is a small killifish, *Fundulus heteroclitus* the mummichog (Fundulidae). Tolerating a wide range of salinities and temperatures, it is found in tidal creeks, coastal bays, salt marshes, and estuaries from Florida to Newfoundland, an exceptionally large latitudinal range for such a small fish (Able, 1984). Now considered two subspecies, the northern and southern subspecies differ in egg morphologies, egg size, spawning substrates, and timing of spawning (Morin & Able, 1983; Marteinsdottir & Able, 1988). More southern populations oviposit in ribbed mussel shells, and more northern populations spawn over vegetation or on muddy substrates. In areas south of Cape Cod, spawning by *F. heteroclitus heteroclitus* occurs around the semilunar high tides.

In Maine, the spawning season is shorter, and *F. heteroclitus macrolepidotus* spawns at many different tidal heights, over a variety of substrates, including sea grass *Spartina*, gravel, and mud (Petersen et al., 2010). The males are brightly colored. Females may lie on one side with the extended abdomen toward a male, or the male and female may swim together with the female in the lead until the male nudges her into a spawning location with a suitable substrate (Petersen et al., 2010). Natural courtship is aquatic.

Oviposition in barnacle or mussel shells or on vegetation allows tidal emergence of the embryos in the upper intertidal zone of estuaries (Taylor & DiMichele, 1983; Taylor, 1999). Adult *F. heteroclitus* can also survive tidal emergence of several hours by breathing air (Halpin & Martin, 1999), although they rarely strand. Regular tidal emergence of the embryos can synchronize environmentally cued hatching (Taylor et al., 1977; DiMichele & Taylor, 1981; Martin, 1999). Development time changes with temperature, as in all ectothermic organisms, and eggs must be submerged to hatch (Taylor, 1999). See Chapter 8 for more on environmentally cued hatching in this species.

The striking semilunar rhythmicity in spawning shown by *F. heteroclitus heteroclitus* has been extensively studied and has been suggested as a model for studying reproductive cycles (Hsaio et al., 1996). However, in Maine, spawning behavior may occur daily with groups of males moving into shallow water with females (Petersen et al., 2010). Laboratory populations of *F. heteroclitus* also spawn daily, particularly the larger individuals, with no semilunar periodicity (Boser et al., 2013). These laboratory populations were taken from Canada, Massachusetts, and Virginia from both subspecies. This plasticity in spawning behavior allows the more northern populations to spawn within a shorter summer season and to depend less on tidal synchrony when those cues are absent.

F. heteroclitus is a model species for scientists working on early development and genetics. Other fundulids have not been as well studied as *F. heteroclitus*, but many species live in similar estuarine habitats and show similar spawning synchrony that peaks around the time of syzygy high tides, as evidenced by the timing of the appearance of eggs. For example, in the Gulf of Mexico, *F. confluentus* (Harrington, 1959), the Gulf killifish *F. grandis* (Greeley & MacGregor, 1983; Hsaio & Meier, 1986, 1989), the longnose killifish *F. similis* (Greeley et al., 1986), the Bayou killifish *F. pulverus,* the diamond killifish *Adinia xenica* (Greeley, 1984), and to a lesser degree the salt marsh topminnow *F. jenkinsi* (Lang et al., 2012) all show semilunar periodicity in their spawning behavior. Because the adult fish move inshore to spawn aquatically in shallow water at high tides, the eggs are tidally emerged occasionally during incubation.

Gobies (Gobiidae) are another family with very successful intertidal and subtidal species. Gobies include some of the most amphibious fishes known, some species that are completely aquatic, and every shading in between. Some gobies produce mucins to form a sperm trail that is placed on a nesting substrate near a female before she spawns, allowing the sperm to be released slowly over time while she places her eggs (Scaggiante et al., 1999).

Most intertidal gobies shelter in water-filled burrows excavated in the mud. On mud flats, particularly during low tides, the burrow water may become extremely hypoxic. As a refuge beneath the surface, these burrows may contain enclosed chambers that are repeatedly filled and refilled with air by the fishes bringing in mouthfuls of fresh air (Ishimatsu et al., 1998; Ishimatsu et al., 2007).

Mudskippers are gobies in the subfamily Oxudercinae. In four genera, there are 5 species of *Boleophthalmus*, 18 species of *Periophthalmus*, 3 species of *Periophthalmodon*, and 4 species of *Scartelaos* (Ishimatsu & Gonzales, 2011). The most amphibious species, in the genus *Periophthalmus*, are frequently emerged even during high tides. Some climb out onto mangrove roots and up into the trees. Species of the less amphibious *Periophthalmodon* genus can emerge but usually remain near the water's edge, with a range of emergence behaviors. The other two genera mostly are aquatic and stay in burrows during low tides.

Eggs of *Ps. argentilineatus* die if forced to remain underwater during development, but must be submerged in order to hatch (Brillet, 1976). *Periphthalmodon schlosseri* and *Periophthalmus modestus* have nest chambers within mud burrows that are filled with air, maintaining a distance between the hypoxic water and the eggs (Ishimatsu et al., 2007, 2009). It is likely that *B. pectinirostris* and *S. histophorus* also place air into their egg chambers. Eggs in the chamber are attached to the upper portion only, in a monolayer (Ishimatsu & Gonzales, 2011).

Males provide parental care to the eggs within the nest. The air within the burrow chamber is replaced periodically with mouthfuls of fresh air by the guarding male in accordance with the tidal cycle, as that affects the duration of aquatic stagnation. Highest delivery of air by the parent to the burrow is during a high tide. Less amphibious species of oxudercine gobies may have similar behavior, but not much is known about these species and some of the data between different studies is conflicting (Ishimatsu & Graham, 2011).

Mudskipper burrows may contain air even when no eggs are present (Ishimatsu et al., 1998). Air is placed in burrows year-round by both male and female fishes of *Boleophthalmus* and *Scartelaos* and may aid the respiration of adult fishes during periods of aquatic hypoxia in low tides, even during nonreproductive periods (Zhang et al., 2000; Lee et al., 2005).

Ishimatsu et al. (2007) speculate that the nest chamber is filled with air before spawning occurs and suggest that both spawning and fertilization take place in an air-filled space on the ceiling of the burrow chamber, an impressive acrobatic feat as well as an unusual strategy for a fish. This awaits confirmation but is intriguing especially in light of the terrestrial spawning of the rockskipper blenny *Andamia tetradactyla* (Shimizu et al., 2006; see Section 3.4).

Other species of *Periophthalmus* and *Boleophthalmus* live in and presumably nest in burrows as well. The ovaries of *Boleophthalmus pectinirostris* follow a semilunar tidal cycle for egg production (Wang et al., 2008), and it is certainly reasonable to hypothesize that both spawning and hatching could be tidally cued in mudskippers (Ishimatsu & Graham, 2011).

Many species of gobids live on mud flats but have not been studied extensively for reproductive behavior. The estuarine goby *Pseudapocryptes elongatus* emerges from water to feed on vegetation on mud flats of India (Das, 1934; Yadav & Singh, 1989). It breathes air, but its reproductive methods have not been studied. It is reported to spawn during the rainy season when the burrows are flooded and to produce large demersal black eggs (Breder & Rosen, 1966). Although it lives in estuaries, it appears to reproduce when the salinity is very low, and its reproduction may have more in common with freshwater fishes than beach-spawning fishes.

In California, the tidewater goby *Eucyclogobius newberryi* nests aquatically in burrows. MacGinitie (1939) reports that they attach the eggs to the walls of the burrow. The male parent fans the eggs, presumably helping to aerate them. Because the species is listed as federally endangered, more recent studies of reproduction have been based in the laboratory, in well-aerated aquariums, and the effects of tidal hypoxia are unknown for this species. The burrows inhabited by *E. newberryi* may be formed by an invertebrate commensal (MacGinitie, 1939). According to Andrew Kinziger (personal communication), tidewater goby eggs are laid in a single layer, so they hang from the ceiling of the artificial PVC tubes on short, flexible stalks. They may sway about with water flow or when nudged by the guarding male. The males make a substantial difference in survival of the eggs. If the male abandons the eggs, they quickly are covered in fungus and few or none hatch. When the male stays with the eggs, fungus problems are reduced and hatch rates are much higher. The males fan with fins and gills at the nest opening and move around the clutch on top of the eggs, apparently nudging and cleaning them.

Also in California, the blind goby *Typhlogobius californiensis* nests in burrows excavated by the ghost shrimp *Neotrypaea californiensis* in sandy and muddy substrates. These burrows may be highly branched with multiple openings that may facilitate water flow over the *T. californiensis* eggs that are laid along the walls (Dumbauld et al., 1996). *T. californiensis* are monogamous, and both parents share the same burrow and care for the eggs (Whiteman & Cote, 2004). Like couples on a double date, pairs of the ghost shrimp live symbiotically with the blind goby pair. The shrimp pump water through the burrow to filter plankton for food. Adult *T. californiensis* are pink in color, indicating vascularization for cutaneous respiration within the aquatic burrows. *T. californiensis* are completely dependent on the currents generated by the shrimp in the burrows, as indicated by their inability to survive or make repairs if the burrow collapses or their shrimp roommates/landlords are removed (MacGinitie, 1939).

The eel goby *Odontamblyopus lacepedii,* a facultative air breather, breathes air by lifting itself up out of its burrow during low tides but is not considered amphibious (Gonzales et al., 2006). Presumably its eggs are fully aquatic. Incubation of eggs in burrows of this species would make a fascinating comparison with the aerial incubation of eggs in the more amphibious mudskippers.

The four-eyed fish *Anableps anableps* (Anablepidae) and its congeners live in estuaries and mate in very shallow water. They differ from oviparous beach-spawning fishes as the female gestates her brood internally, releasing fully formed juveniles in several weeks (Oliveira et al., 2011). These fish are considered intertidal spawners (Helfman et al., 2009), so calling this reproductive behavior "beach spawning" seems appropriate.

3.6   SUMMARY OF BEACH SPAWNING BY RESIDENT INTERTIDAL FISHES

Within all this variety, some patterns may emerge. In general, resident intertidal fishes spawn in pairs, even though rarely monogamous. They do not form large aggregations for group spawning. Clutches are placed in fixed sites, whether guarded or not, rather than broadcast.

Males, females, or both may guard clutches, or there may be no parental care. Care of eggs involves cleaning and aerating them and probably defends against infection and predation. Some cannibalism of conspecific eggs may occur during spawning or nest guarding. If males guard, the presence of eggs may attract other females and allow him to obtain multiple matings and clutches.

Different lineages differ; reproduction in intertidal blennies is similar to subtidal forms, with lots of complex courtship and parental care. Among the many species of sculpins, some show copulation with no commitment; others IGA or external fertilization. Nest parasitism occurs on top of the clutches of one of the few sculpins that actually guards its nests (Kent et al., 2011). It is intriguing that of all the sculpins that reproduce in the intertidal zone, the only two that show parental care are species that usually live subtidally, *Enophrys bison* and *Clinocottus acuticeps* (DeMartini, 1978; Kent et al., 2011; Table 3.1; see Chapter 4 for more details).

Stichaeids and zoarcids provide parental care by wrapping around a spherical clutch in the space beneath a boulder, while sculpins oviposit in a monolayer on the ceiling of the chamber under a boulder. Positioning eggs on ceilings of the chamber exposes them to air rather than leaving them in hypoxic water during low tides. Parents fan with fins to aerate eggs and increase water flow in burrows, and this also aids in keeping the egg surfaces clean. A monolayer of eggs seems to be best for clutches placed in crevices or under boulders. Spawning by clingfishes and blennies is done only in areas that do not already have eggs, although this is not the case for sculpins. If eggs are covered by another layer of eggs, the developmental rate of the deeper set slows.

For burrow nesters, cleaning and caring for the eggs may involve one but often both parents. Parental care is probably necessary for clutches in mud burrows to provide aeration and prevent infection. On the other hand, use of empty shells for nesting by *Fundulus* does not require parental care, nor does attachment of eggs to vegetation.

Substrate for spawning is frequently or constantly available to resident intertidal fishes, but environmental cues such as semilunar high tides may be necessary for behaviors associated with courtship and spawning, as well as external fertilization. Some species that otherwise show synchrony with semilunar tides can spawn more frequently in the laboratory under constant conditions that are conducive to oviposition.

Rocky intertidal substrates are more likely to house parental care and nests for intertidal fishes than are sandy or gravel beaches. No resident intertidal fishes spawn on sandy or gravel beaches, but many different fishes that live subtidally most of the time make spawning migrations into the intertidal zone to spawn in gravel or sandy beaches. These species are described next, in Chapter 4.

REFERENCES

Abel, E. F. (1973). Zur Oko-Ethologie des amphibisch lebended Fisches *Alticus saliens* (Forster) und von *Entomacrodus vermiculatus* (Val.) (Blennioidea, Salariidae), unter besonderer Berucksichtigung des Forpflanzungsverhaltens. *Sitzungsberichte der Osterreichischen Akademie der Wissenschaften* 183, 137–153.

Able, K. W. (1984). Variation in spawning site selection of the mummichog, *Fundulus heteroclitus*. *Copeia* 1984, 522–525. http://www.jstor.org/stable/1445207.

Almada, V. C. & Santos, R. S. (1995). Parental care in the rocky littoral: adaptation and exaptation in Atlantic and Mediterranean blennies. *Reviews in Fish Biology and Fisheries* 5, 23–37.

Aryafar, H. (2012). *Sexual Dimorphism in Reproductive Morphology of Leuresthes tenuis (Atherinopsidae).* Master's thesis. California State University, Fullerton.

Barata, E. N., Serrano, R. M., Miranda, A., Nogueira, R., Hubbard, P. C. & Canario, A. V. M. (2008). Putative pheromones from the anal glands of male blennies attract females and enhance male reproductive success. *Animal Behaviour* 75, 379–389.

Blaxter, J. H. S. (1988). Pattern and variety in development. In *Fish Physiology, Volume 8* (Hoar, W. H., Randall, D. J. & Conte, F. P., eds.), 1–59. San Diego: Academic Press.

Boser, T., Munkittrick, K., Nacci, D. E. & MacLatchy, D. L. (2013). Laboratory spawning patterns of mummichogs, *Fundulus heteroclitus* (Cyprinodontiformes: Fundulidae). *Copeia* 2013, 527–538. DOI: 10.1643/CI-11-175.

Brantley, R. K. & Bass, A. H. (1994). Alternative male spawning tactics and acoustic signaling in the plainfin midshipman fish, *Porichthys notatus*. *Ethology* 96, 213–232.

Breder, C. M. & Rosen, D. E. (1966). *Modes of Reproduction in Fishes.* Garden City: Natural History Press, 941 pp.

Brillet, C. (1976). Structure du terrier, reproduction et comportment des jeunes chez le poisson amphibie *Periophthalmus sobrinus* Eggert. *Revue d'Ecologie: La Terre et la Vie* 30, 465–483.

Brown, C. R., Gordon, M. S. & Martin, K. L. M. (1992). Aerial and aquatic oxygen uptake in the amphibious Red Sea rockskipper fish, *Alticus kirki* (Family Blenniidae). *Copeia* 1992, 1007–1013.

Clutton-Brock, T. H. (1991). *The Evolution of Parental Care.* Princeton, NJ: Princeton University Press, 352 pp.

Coleman, R. M. (1992). Reproductive biology and female parental care in the cockscomb prickleback, *Anoplarchus purpurescens* (Pisces: Stichaeidae). *Environmental Biology of Fishes* 35, 177–186.

Coleman, R. M. (1999). Parental care in intertidal fishes. In *Intertidal Fishes: Life in Two Worlds* (Horn, M. H., Martin, K. L. M. & Chotkowski, M. A., eds.), 165–180. San Diego: Academic Press.

Das, B. K. (1934). The habits and structure of *Pseudapocryptes lanceolatus*, a fish in the first stages of structural adaptation to aerial respiration. *Proceedings of the Royal Society of London* 115, 422–431.

DeMartini, E. E. (1978.) Spatial aspects of reproduction in buffalo sculpin, *E. bison*. *Environmental Biology of Fishes* 3, 331–336.

DeMartini, E. E. (1999). Intertidal spawning. In *Intertidal Fishes: Life in Two Worlds* (Horn, M. H., Martin, K. L. M. & Chotkowski, M. A., eds.), 143–164. San Diego: Academic Press.

DiMichele, L. & Taylor, M. (1981). The mechanism of hatching in *Fundulus heteroclitus*: Development and physiology. *Journal of Experimental Zoology* 217, 73–79.

Dumbauld, B. R., Armstrong, D. A. & Feldman, K. L. (1996). Life-history characteristics of two sympatric thalassinidean shrimps, *Neotrypaea californiensis* and *Upogebia pugettensis*, with implications for oyster culture. *Journal of Crustacean Biology* 16, 689–708.

Eschmeyer, W. N., Herald, E. S. & Hammann, H. (1983). *A Field Guide to Pacific Coast Fishes of North America*. Peterson Field Guide Series. Boston: Houghton Mifflin.

Ferry-Graham, L. A., Drazen, J. C. & Franklin, V. (2007). Laboratory observations of reproduction in the deep-water Zoarcids *Lycodes cortezianus* and *Lycodapus mandibularis* (Teleostei: Zoarcidae). *Pacific Science* 61, 129–139. DOI: http://dx.doi.org/10.1353/psc.2007.0004.

Giacomello, E., Marchini, D. & Rasotto, M. B. (2006). A male sexually dimorphic trait provides antimicrobials to eggs in blenny fish. *Biology Letters* 2, 330–333.

Giacomello, E., Neat, F. C. & Rasotto, M. B. (2008). Mechanisms enabling sperm economy in blenniid fishes. *Behavioral Ecology and Sociobiology* 62, 671–680.

Gibson, R. N. (1982). Recent studies on the biology of intertidal fishes. *Oceanography and Marine Biology Annual Reviews* 20, 363–414.

Goncalves, E. J. & Almada, V. C. (1998). A comparative study of territoriality in intertidal and subtidal blennioids (Teleostei, Blennioidei). *Environmental Biology of Fishes* 51, 257–264.

Gonzales, T. T., Katoh, M. & Ishimatsu, A. (2006). Air breathing of aquatic burrow-dwelling eel goby, *Odontamblyopus lacepedii* (Gobiidae: Amblyopinae). *Journal of Experimental Biology* 209, 1085–1092.

Gosztonyi, A. E. (1977). Results of the research cruises of FRV "Walter Herwig" to South America. XLVIII: Revision of South American Zoarcidae (Osteichthyes, Blennioidei) with descriptions of three new genera and five new species. *Arbeit Fish Wissenschaft* 27, 191–249.

Greeley, M. S., Jr. (1984). Spawning by *Fundulus pulverous* and *Adinia xenica* (Cyprinodontidae) along the Alabama coast associated with semilunar tidal cycles. *Copeia* 1984, 797–800.

Greeley Jr., M. S. & MacGregor, R. (1983). Annual and semilunar reproductive cycles of the gulf killifish, *Fundulus grandis*, on the Alabama gulf coast. *Copeia* 1983, 711–718.

Greeley Jr., M. S., Marion, K. R. & MacGregor III, R. (1986). Semilunar spawning cycles of *Fundulus similis* (Cyprinodontidae). *Environmental Biology of Fishes* 17, 125–131.

Halpin, P. M. & Martin, K. L. M. (1999). Aerial respiration in the salt marsh fish *Fundulus heteroclitus* (Fundulidae). *Copeia* 1999, 743–748.

Harrington, R. W., Jr. (1959). Effect of four combinations of temperature and daylength on the oogenic cycle of a low latitude fish, *Fundulus confluentes*. *Zoologica* 44, 149–168.

Hastings, P. A. & Petersen, C. W. (2010). Parental care, oviposition sites, and mating systems of Blennoid fishes. In *Reproduction and Sexuality in Marine Fishes* (Cole, K. S., ed.), 91–116. Berkeley: University of California Press.

Hayakawa, Y., Akiyama, R. & Munehara, H. (2004). Antidispersive effect induced by parasperm contained in semen of a cottid fish, *Hemilepidotus gilberti*: estimation by models and experiments. *Japanese Journal of Ichthyology* 51, 31–42.

Hayakawa, Y., Komaru, A. & Munehara, H. (2002). Obstructive role of the dimorphic sperm in a non-copulatory marine sculpin, *Hemilepidotus gilberti*, to prevent other males' eusperm from fertilization. *Environmental Biology of Fishes* 64, 419–427.

Helfman, G., Collette, B. B., Facey, D. E. & Bowen, B. W. (2009). *The Diversity of Fishes: Biology, Evolution, and Ecology*. Oxford, UK: Wiley-Blackwell.

Horn, M. H. & Riegle, K. C. (1981). Evaporative water loss and intertidal vertical distribution in relation to body size and morphology of stichaeoid fishes from California. *Journal of Experimental Marine Biology and Ecology* 50, 273–288.

Horn, M. H., Martin, K. L. M. & Chotkowski, M. A. (Eds.) (1999). *Intertidal Fishes: Life in Two Worlds*. San Diego: Academic Press.

Hsaio, S.-M. & Meier, A. H. (1986). Spawning cycles of the Gulf killifish, *Fundulus grandis*, in closed circulation systems. *Journal of Experimental Zoology* 240, 105–112.

Hsaio, S.-M. & Meier, A. H. (1989). Comparison of semilunar cycles of spawning activity in *Fundulus grandis* and *F. heteroclitus* held under constant laboratory conditions. *Journal of Experimental Zoology* 252, 213–218.

Hsaio, S.-M., Limesand, S. W. & Wallace, R. A. (1996). Semilunar follicular cycle of an intertidal fish: the *Fundulus* model. *Biology of Reproduction* 54, 809–818.

Hubbs, C. (1966). Fertilization, initiation of cleavage, and developmental temperature tolerance of the cottid fish, *Clinocottus analis*. *Copeia* 1966, 29–42.

Hughes, G. W. (1985). The comparative ecology and evidence for resource partitioning in two pholidid fishes (Pisces: Pholididae) from southern British Columbia eelgrass beds. *Canadian Journal of Zoology* 63, 76–85.

Ishimatsu, A. & Gonzales, T. T. (2011). Mudskippers: front runners in the modern invasion of land. In *The Biology of Gobies* (Patzner, R. A., Van Tassell, J. L., Kovacic, M. & Kapoor, B. G., eds.), 609–638. Enfield, NH: CRC Press, Taylor & Francis.

Ishimatsu, A. & Graham, J. B. (2011). Roles of environmental cues for embryonic incubation and hatching in mudskippers. *Integrative and Comparative Biology* 51, 38–48. DOI: 10.1093/icb/icr018.

Ishimatsu, A., Hishida, Y., Takita, T., Kanda, T., Oidawa, S., Takeda, T. & Khoo, K. H. (1998). Mudskippers store air in their burrows. *Nature* 391, 237–238.

Ishimatsu, A., Takeda, T., Tsuhako, Y., Gonzales, T. T. & Khoo, K. H. (2009). Direct evidence for aerial egg deposition in the burrows of the Malaysian mudskipper *Periophthalmodon schlosseri*. *Ichthyological Research* 56, 417–420.

Ishimatsu, A., Yoshida, Y., Itoki, N., Takeda, T., Lee, H. J. & Graham, J. B. (2007). Mudskippers brood their eggs in air but submerge them for hatching. *Journal of Experimental Biology* 210, 3946–3954.

Kent, D. I., Fisher, J. D. & Marliave, J. B. (2011). Interspecific nesting in marine fishes: spawning of the spinynose sculpin, *Asemichthys taylori*, on the eggs of the buffalo sculpin, *Enophrys bison*. *Ichthyological Research* 58, 355–359.

Lang, E. T., Brown-Peterson, N. J., Peterson, M. S. & Slack, W. T. (2012). Seasonal and tidally driven patterns in the saltmarsh topminnow, *Fundulus jenkinsi*. *Copeia* 2012, 451–459.

Lee, H. J., Martinez, C. A., Hertzberg, K. J., Hamilton, A. L. & Graham, J. B. (2005). Burrow air phase maintenance and respiration by the mudskipper *Scartelaos histophorus* (Gobiidae: Oxudercinae). *Journal of Experimental Biology* 208, 169–177.

Little, C. (2000). *The Biology of Soft Shores and Estuaries*. Oxford, UK: Oxford University Press, 270 pp.

MacGinitie, G. E. (1939). The natural history of the blind goby, *Typhlogobius californiensis* Steindachner. *American Midland Naturalist* 21, 489–505.

Marliave, J. B. (1977). Substratum preferences of settling larvae of marine fishes reared in the laboratory. *Journal of Experimental Marine Biology and Ecology* 27, 47–60.

Marliave, J. B. & DeMartini, E. E. (1977). Parental behavior of intertidal fishes of the stichaeid genus *Xiphister*. *Canadian Journal of Zoology* 55, 60–63.

Marteinsdottir, G. & Able, K. W. (1988). Geographic variation in egg size among populations of the mummichog, *Fundulus heteroclitus* (Pisces: Fundulidae). *Copeia* 1988, 471–478.

Martin, K. L. M. (1993). Aerial release of CO_2 and respiratory exchange ratio in intertidal fishes out of water. *Environmental Biology of Fishes* 37, 189–196.

Martin, K. L. M. (1999). Ready and waiting: delayed hatching and extended incubation of anamniotic vertebrate terrestrial eggs. *American Zoologist* 39, 279–288.

Martin, K. L. M. & Bridges, C. R. (1999). Respiration in water and air. In *Intertidal Fishes: Life in Two Worlds* (Horn, M. H., Martin, K. L. M. & Chotkowski, M. A., eds.), 54–78. San Diego: Academic Press.

Martin, K. L. M., Bailey, K., Moravek, C. & Carlson, K. (2011). Taking the plunge: California grunion embryos emerge rapidly with environmentally cued hatching. *Integrative and Comparative Biology* 51, 26–37. DOI: 10.1093/icb/icr037.

Martin, K. L. M., Van Winkle, R. C., Drais, J. E. & Lakisic, H. (2004). Beach spawning fishes, terrestrial eggs, and air breathing. *Physiological & Biochemical Zoology* 77, 750–759.

Matallanas, J., Rucabado, J., Lloris, D. & Pilar Olivar, M. (1990). Early stages of development and reproductive biology of the South American eelpout *Austrolycus depressiceps* Regan, 1913 (Teleostei: Zoarcidae). *Sciencias Marinas* 54, 257–261.

Miki, T., Yoshida, H. & Amaoka, K. (1987). Rare stichaeid fish, *Pseudalectrias tarasovi* (Popov), from Japan and its larvae and juveniles. *Bulletion of the Faculty of Fisheries Hokkaido University* 38, 1–13.

Morin, R. P. & Able, K. W. (1983). Patterns of geographic variation in the egg morphology of the fundulid fish, *Fundulus heteroclitus*. *Copeia* 1983, 726–740.

Morris, R. W. (1956). Clasping mechanism of the cottid fish *Oligocottus snyderi* Greeley. *Pacific Science* 10, 314–317.

Munday, P. L., Kuwamura, T. & Kroon, F. J. (2010). Bidirectional sex change in marine fishes. In *Reproduction and Sexuality in Marine Fishes* (Cole, K. S., ed.), 241–271. Berkeley, CA: University of California Press.

Munehara, H., Takano, K. & Koya, Y. (1989). Internal gamete association and external fertilization in the elkhorn sculpin, *Alcichthys alcicornis. Copeia* 1989, 673–678.

Munehara, H., Takano, K. & Koya, Y. (1991). The little dragon sculpin *Blepsias cirrhosis*, another case of internal gamete association and external fertilization. *Japanese Journal of Ichthyology* 37, 391–394.

Munehara, H., Koya, Y., Hayakawa, Y. & Takano, K. (1997). Extracellular environments for the initiation of external fertilization and micropylar plug formation in a cottid species, *Hemitripterus villosus* (Pallas) (Scorpaeniformes) with internal insemination. *Journal of Experimental Marine Biology and Ecology* 211, 279–289.

Munoz, M. (2010). Reproduction in Scorpaeniformes. In *Reproduction and Sexuality in Marine Fishes* (Cole, K.S., ed.), 65–116. Berkeley, CA: University of California Press.

Neat, F. C. (2001). Male parasitic spawning in two species of triplefin blenny (Trypterigiidae): contrasts in demography, behaviour and gonadal characteristics. *Environmental Biology of Fishes* 61, 57–64. DOI: 10.1023/A:1011074716758.

Oliveira, V. deA., Fontoura, N. F. & Montag, L. F. deA. (2011). Reproductive characteristics and the weight-length relationship in *Anableps anableps* (Linnaeus, 1758) (Cyprinodontiformes: Anablepidae) from the Amazon Estuary. *Neotropical Ichthyology* 9, 757–766.

Petersen, C. W. (1989). Females prefer mating males in the carmine triplefin, *Axoclinus carminalis*, a paternal brood guarder. *Environmental Biology of Fishes* 26, 213–221.

Petersen, C. W. & Hess, H. C. (2011). Evolution of parental behavior, egg size, and egg mass structure in sculpins. In *Adaptation and Evolution in Cottoid Fishes* (Goto, A., Munehara, H. & Yabe, M., eds.), 194–203. Tokai University Press, Kanagawa, Japan.

Petersen, C. W., Mazzoldi, C., Zarrella, K. A. & Hale, R. E. (2005). Fertilization mode, sperm characteristics, mate choice and parental care patterns in *Artedius* spp. (Cottidae). *Journal of Fish Biology* 67, 239–254. DOI: 10.111/j.1095-8649.2005.00732.x.

Petersen, C. W., Salinas, S., Preston, R. L. & Kidder III, G. W. (2010). Spawning periodicity and reproductive behavior of *Fundulus heteroclitus* in a New England salt marsh. *Copeia* 2010, 203–210. DOI: http://dx.doi.org/10.1643/CP-08-229.

Petersen, C. W., Zarrella, K. A., Ruben, C. A. & Mazzoldi, C. (2004). Reproductive biology of the rosylip sculpin, an intertidal spawner. *Journal of Fish Biology* 64, 863–875.

Phillips, R. R. (1977). Behavioural field study of the Hawaiian rockskipper, *Istiblennius zebra* (Teleostei, Blenniidae). I. Ethogram. *Zeitschrift fur Tierpsychologie* 43, 1–22.

Pierce, B. E. & Pierson, K. B. (1990). Growth and reproduction of the tide pool sculpin *Oligocottus maculosus*. *Japanese Journal of Ichthyology* 36, 410–418.

Pires, T. H. S. & Gibran, F. Z. (2011). Intertidal life: field observations on the clingfish *Gobiesox barbatulus* in southeastern Brazil. *Neotropical Ichthyology* 9, 233–240.

Qasim, S. Z. (1957). The biology of *Centronotus gunnellus* (L.) (Teleostei). *Journal of Animal Ecology* 26, 389–401.

Ragland, H. C. & Fischer, E. A. (1987). Internal fertilization and male parental care in the scalyhead sculpin, *Artedius harringtoni. Copeia* 1987, 1059–1062.

Rizzo, E., Sato, Y., Barreto, B. P. & Godinho, H. P. (2002). Adhesiveness and surface patterns of eggs in neotropical freshwater teleosts. *Journal of Fish Biology* 61, 615–632. DOI: 10.1111/j.1095=6849.2002.tb00900.x.

Rodgers, E. W., Earley, R. I. & Grober, M. S. (2007). Social status determines sexual pheno-type in the bi-directional sex changing blueband goby *Lythrypnus dalli. Journal of Fish Biology* 70, 1660–1668.

Scaggiante, M., Mazzoldi, C., Petersen, C. W. & Rasotto, M. B. (1999). Sperm competition and mode of fertilization in the grass goby *Zosterisessor ophiocephalus* (Teleostei: Gobiidae). *Journal of Experimental Zoology* 283, 81–90.

Shimizu, N., Sakai, Y., Hashimoto, H. & Gushima, K. (2006). Terrestrial reproduction by the air-breathing fish *Andamia tetradactyla* (Pisces: Blenniidae) on supralittoral reefs. *Journal of Zoology* 269, 357–364.

Shinohara, G., Nazarkin, M. V. & Chereschnev, I. A. (2004). *Magadania skopetsi*, a new genus and species of Zooarcidae (Teleostei: Perciformes) from the Sea of Okhotsk. *Ichthyological Research* 51, 137–145. DOI: 10.1007/s10228-004-2-0209-7.

Stephens, J. S., Johnson, R. K., Key, G. S. & McCosker, J. E. (1970). The comparative ecol-ogy of three sympatric species of California blennies of the genus *Hypsoblennius* Gill (Teleostomi, Blenniidae). *Ecological Monographs* 40, 213–233.

Strathmann, R. R. & Hess, H. C. (1999). Two designs of marine egg masses and their divergent consequences for oxygen supply and desiccation in air. *American Zoologist* 39, 253–260.

Taylor, M. H. (1984). Lunar synchronization of fish reproduction. *Transactions of the American Fisheries Society* 113, 484–493.

Taylor, M. H. (1999). A suite of adaptations for intertidal spawning. *American Zoologist* 39, 313–320.

Taylor, M. H. & DiMichele, L. (1983). Spawning site utilization in a Delaware population of *Fundulus heteroclitus* (Pisces: Cyprinodontidae). *Copeia* 1983, 719–725.

Taylor, M. H., DiMichele, L. & Leach. G. J. (1977). Egg stranding in the life cycle of the mummichog, *Fundulus heteroclitus. Copeia* 1977, 397–399.

Tewksbury, H. T. & Conover, D. O. (1987). Adaptive significance of intertidal egg deposition in the Atlantic silverside *Menidia menidia. Copeia* 1987, 76–83.

Wang, Q., Hong, W., Chen, S. & Zhang, Q. (2008). Variation with semilunar periodicity of plasma steroid hormone production in the mudskipper *Boleophthalmus pectinirostris. General & Comparative Endocrinology* 155, 821–826. DOI: 10.1016/j.ygcen.2007.10.008.

Warner, R. R. (1988). Sex change and the size-advantage model. *Trends in Ecology and Evolution* 3, 133–136.

Whiteman, E. A. & Cote, I. M. (2004). Monogamy in marine fishes. *Biological Reviews* 79, 351–375.

Wirtz, P. (1978). The behaviour of the Mediterranean *Tripterygion* species (Pisces, Blennioidei). *Zeitschrift fur Tierpsychologie* 48, 142–174.

Yadav, A. N. & Singh, B. R. (1989). Gross structure and dimensions of the gill in an air-breathing estuarine goby, *Pseudapocryptes lanceolatus. Japanese Journal of Ichthyology* 36, 252–259.

Yamahira, K. (1996). The role of intertidal egg deposition on survival of the puffer, *Takifugu niphobles* (Jordan et Snyder), embryos. *Journal of Experimental Marine Biology and Ecology* 198, 291–306.

Zander, C. D. (1972). Beziehungen zwischen Korperbau und Lebensweise bei Blenniidae (Pisces) aus dem Roten Meer. I. Aussere Morphologie. *Marine Biology* 13, 238–246.

Zander, C. D., Nieder, J. & Martin, K. (1999). Vertical distribution patterns. In *Intertidal Fishes: Life in Two Worlds* (Horn, M. H., Martin, K. L. M. & Chotkowski, M. A., eds.), 26–53. San Diego: Academic Press.

Zhang, J., Taniguchi, T., Takita, T. & Ali, A. B. (2000). On the epidermal structure of *Boleophthalmus* and *Scartelaos* mudskippers with reference to their adaptation to terrestrial life. *Ichthyological Research* 47, 359–366.

4 Vacation Sex: Subtidal Fishes That Make Spawning Migrations to the Beach

Many species of subtidal fishes are found transiently in shallow water in the high intertidal zone, typically during a high tide (Gibson, 1982). Subtidal fishes that spawn on beaches in the intertidal zone come from a variety of different lineages, including smelts (Osmeridae), New World silversides (Atherinopsidae), herring (Clupeidae), toadfish (Batrachoididae), fugu puffer (Tetraodontidae), sandfish (Trichodontidae), sand lance (Ammodytidae), and even flatfish (Pleuronectidae). In most cases the spawning is timed with the higher tide levels or a higher shore water level caused by large wind waves (DeMartini, 1999). In all cases spawning in each species is associated with a particular type of substrate for the eggs, whether sand, rock, or vegetation, that is submerged and becomes available for oviposition with the highest tides (Conover & Kynard, 1984). As for the resident intertidal fishes, in general the eggs deposited by subtidal migrants in the upper beach are exposed to air for some or all of their incubation period (Martin et al., 2004).

Many of these subtidal species of fishes form large aggregations near a beach-spawning area at the proper time. A spawning migration may be a spectacular parade of thousands of fishes, attracting hordes of predators and anglers, as is the case with the beach-spawning capelin *Mallotus villosus*, or it may be so subtle that it is rarely observed, as in the Pacific sand lance *Ammodytes hexapterus*.

Spawning may take place on beaches made of rocks, sand, or gravel. Some species require vegetation or other substrates for oviposition. In most beach-spawning fishes, the migrations are relatively short and the aggregations may last only a few hours or days. Spawning may take place between pairs, or in groups of one or more females and multiple males. Most beach-spawning fishes are able to reproduce repeatedly, both during one season and across several years. Other species die after reproducing once. See Chapter 6 for some of the risks that go along with the rewards of spawning in a group.

Some territorial coral reef species, living in a relatively constant environment with the substrate for nests always available, spawn every day. Similarly, if the appropriate substrate along with plenty of food is available, beach-spawning *Menidia menidia* can spawn daily (Conover & Kynard, 1984). Likewise, the estuarine *Fundulus heteroclitus* spawns only at semilunar tides in the Delaware portion of its range (Taylor, 1984), but more frequently during the shorter summers in a shallow bay in Maine (Petersen et al., 2010), as discussed in Chapter 3.

Unlike the resident intertidal fishes, this group shows little parental care after oviposition, although there are exceptions (Coleman, 1999). This chapter summarizes information for fishes that migrate from subtidal or other waters to spawn on beaches (Table 4.1), and Table 4.2 shows some of the variety of mating systems seen in these fishes.

TABLE 4.1
Subtidal Beach Spawning Species

Family	Genus, Species	Location of Study Population	References
Atherinopsidae	Colpichthys regis	Gulf of California, Mexico	Russell et al., 1989
	Menidia menidia	Atlantic coast of North America	Middaugh, 1981
Clupeidae	Clupea pallasii pallasii	Pacific coast of North America	Jones, 1972
Cottidae	Clinocottus acuticeps	British Columbia, Canada	Marliave, 1981
	Enophrys bison	Puget Sound, British Columbia	Kent et al., 2011
Galaxiidae	Galaxias maculatus	New Zealand	MacDowall, 1968
Gasterosteidae	Gasterosteus aculeatus	Canada	MacDonald et al., 1996
Fundulidae	Fundulus heteroclitus	Delaware Bay, Atlantic, USA	Taylor, 1999
	Fundulus grandis	Gulf of Mexico, USA	Martinez et al., 2006
Osmeridae	Hypomesus pretiosus	Pacific coast of North America	Penttila et al., 2007
	Hypomesus japonicas	Japan	Hirose & Kawaguchi, 1998
	Mallotus villosus	Canada, Iceland, Alaska	Nakashima & Wheeler, 2002
	Spirinchus starksi	Pacific Northwest, Canada	Penttila, 2001
Tetraodontidae	Takifugu niphobles	Japan	Yamahira, 1996

Note: These beach spawning species live in the subtidal zone. They remain aquatic while spawning and leave the intertidal zone soon after. Spawning occurs at the water's edge during a high tide, and this allows the eggs to be emerged into air with subsequent low tides.

TABLE 4.2
Mating Systems Seen in Fishes that Migrate to Spawn on Beaches

Type, Brief Description	Example Species
Copulation (may be internal fertilization or internal gamete association)	
With male egg guarding	Clinocottus acuticeps
With egg guarding, nest parasitism	Enophrys bison
Polygyny, territorial males guard multiple clutches	
With "sneaker" males	Porichthys notatus
Pairs and polyandry, multiple paternity	Leuresthes tenuis, Mallotus villosus, Takifugu niphobles

Note: See text for additional details on each species.

4.1   SOME SPECIES MIGRATE INTO THE ROCKY INTERTIDAL ZONE TO SPAWN

The family Cottidae, a very diverse clade, contains many species of sculpins that live and spawn in the rocky intertidal zone (see Section 3.4). Other cottid species live and spawn subtidally. However in one unusual case, a subtidal cottid, the sharpnose sculpin *Clinocottus acuticeps*, lives in the shallow subtidal zone but migrates high up in the intertidal zone for spawning (Marliave, 1981). This sculpin deposits its eggs among the fronds of the rockweed *Fucus distichus* and may remain nearby, guarding the eggs while the clutch is submerged during high tides. But when the tides ebb and the eggs are stranded in air, the parent usually swims away, to remain in the water (Marliave, 1981). The eggs with their developing embryos are typically covered by seaweed, which protects them from drying out and from overheating. If the seaweed is removed experimentally, the embryos perish from exposure (Marliave, 1981).

Cottids are very diverse, with numerous intertidal and subtidal species. Only two subtidal species place their eggs in the intertidal zone, *C. acuticeps* and occasionally *Enophrys bison*. The buffalo sculpin, *Enophrys bison*, lives near shore and nests subtidally in Alaska and Canada. In Puget Sound, *E. bison* nests in the intertidal zone (DeMartini, 1978). A guarding male provides care by fanning the nest containing eggs produced by several females. A sympatric subtidal sculpin, *Asemichthys taylori*, oviposits over existing clutches of *E. bison* eggs in British Columbia (Kent et al., 2011). These eggs are also guarded by the *E. bison* male and hatch sooner than the *E. bison* eggs. Covering the older eggs of *E. bison* appears to slow their development, and this is considered a form of nest parasitism previously unknown among fishes (Kent et al., 2011). It is intriguing that none of the intertidal cottids remain guarding nests throughout incubation, but both of these subtidal species that spawn intertidally show parental care of the nests that are outside of their adult habitat (Petersen & Hess, 2011).

The plainfin midshipman *Porichthys notatus* (Batrachoididae) is the only toadfish species that spawns on beaches. Its common name refers to the double row of photophores on its belly that resemble the arrangement of brass buttons on a naval uniform. Although usually found deep in subtidal waters, larger males move onshore at the start of the spawning season to set up nesting territories in small pools under intertidal boulders (Arora, 1948). Males apparently breed subtidally also, to a depth of 81 m (Feder et al., 1974), but little is known about their subtidal reproductive activities.

Male *P. notatus* create a nesting chamber between the base of the boulder and the substrate by body and fin movements to form a shallow depression where water can collect during low tides. The better the boulder, the more dominant the male in that territory (DeMartini, 1988). The male then attracts the desired females by "humming," vibrating the swim bladder to produce an irresistible call that is amplified by the resonance of the nest chamber, loud enough to be heard by nearby humans (Vasconcelas et al., 2011). The smaller female approaches underwater during a high tide, and if she is suitably impressed, she attaches some of her eggs on the ceiling of the chamber with tiny stalks like so many party balloons (Love, 2011). Oviposition on the underside of a boulder is similar to the type seen in many resident intertidal fishes (Coleman, 1999). See Figure 4.1.

P. notatus has two forms of males (Brantley & Bass, 1994), with alternative reproductive tactics (ART). The territorial male (Type I) prepares his nesting chamber by cleaning and expanding it. Then he begins calling to attract females. When he starts to fertilize her eggs, another smaller male (Type II) may also enter the nest along with the female. Territorial males tend to behave aggressively against other males, but the "sneaker" male looks so much like a female that he does not incite this response. The Type II males can surreptitiously provide milt to the newly present clutch, fathering at least some of the offspring. The dimorphism in the two types of males is profound and extends to the cellular structure of the swim bladder. Muscle cells of the Type I males are 40 percent larger and contain more, larger myofibrils than the Type II males or females (Bass & Marchaterre, 1989). Toadfish produce sound by vibrating the hollow, gas-filled swim bladder like a drum, using muscles that have the fastest contractions of any known vertebrate muscles (Fine et al., 2001).

(a)

(b)

FIGURE 4.1   Beach-spawning fishes produce demersal eggs that may (a) be scattered over a mobile substrate to which they adhere, as for *Takifugu niphobles*; (b) adhere to kelp or other vegetation with filaments, as for *Clupea pallasi pallasii*. (Artwork by Andrea Lim.)

(c)

(d)

FIGURE 4.1 (*Continued*) Beach-spawning fishes produce demersal eggs that may (c) be placed directly onto a fixed substrate to which they adhere, as for *Porichthys notatus* (the eggs are on the underside of the boulder that has been lifted up); or (d) be buried in sand with nonadherent eggs, as in *Leuresthes tenuis*. (Photos by Karen Martin.)

The females and Type II males then return to deeper water, while the resident Type I male remains in the intertidal zone with the developing eggs, caring for them up to and even after hatching (Crane, 1981).

The eggs are attached individually by stalks to the ceiling of the chamber, the underside of the boulder. Upon hatching, the prolarvae remain tethered to the spot for several weeks. The male guards the eggs and larvae without ever leaving, becoming more emaciated over time and worse for the wear as his skin may be injured by crabs or other predators (Feder et al., 1974). His testosterone levels are high when he is courting and beginning to obtain eggs for the nest, but over the course of their development his hormone levels drop (Knapp et al., 1999). It is possible that the challenges of child rearing lead to an early grave for many of these dedicated fathers.

Its congener *P. myriaster*, the specklefin midshipman, replaces *P. notatus* in southern California and Mexico, living in shallow water and bays over mud or sand substrates. *P. myriaster* produces somewhat smaller eggs during reproduction that are "similar to plainfin midshipmen" (Feder et al., 1974, p. 112), but this species has been studied far less than *P. notatus* and its reproductive pattern has not been described in any detail. An obvious question would be to examine the reproductive patterns of subtidally spawning *Porichthys* of either species to see if males there hum and guard nests too, or if that only occurs in the intertidal zone.

Some spawning migrations are so subtle that they have not been observed but are inferred from nest sites. The Pacific sandfish *Trichodon trichodon* (Trichodontidae) spends most of its larval and juvenile life buried in sandy or gravel substrates near shore in shallow water (Bailey et al., 1983), although adults are found in deeper water (Eschmeyer et al., 1983). The adult *T. trichodon* may migrate along shore from offshore sandy habitats to find suitable spawning habitat in the rocky intertidal zone. Its eggs have been found as gelatinous masses exposed to air by tides in surge channels of rocky intertidal zones, which are some distance from sandy beach habitats (Marliave, 1980), where they are kept wet by wave action. Eggs of *T. trichodon* take up to a year to hatch in the cool waters of Alaska and Canada, and no parental care is known for this species, nor are any details about the spawning behaviors or information about whether this forage fish species may spawn in other habitats as well. Their young school immediately with their conspecifics and also with other forage fishes including pink salmon *Oncorhynchus gorbuscha,* chum salmon *O. keta*, cod *Gadus morhua*, Pacific herring *Clupea pallasi pallasi*, and walleye pollock *Theragra chalcogramma* (Bailey et al., 1983; Thedinga et al., 2006).

4.2 SOME SPECIES MIGRATE ONTO GRAVEL BEACHES TO SPAWN

Gravel beaches have coarse sediments rather than the solid substrate of rocky tide pools, benches, and boulders. They are found in temperate and polar coasts. Gravel beaches drain well but water is held interstitially, providing high humidity with rapid diffusion of oxygen through the air spaces. The sediment provides some shade for the embryos and attachment sites for adherent eggs. On gravel beaches, some eggs attach to the edges rather than the flat sides of the gravel; this exposes more of the egg surface to the surrounding water or air.

The true smelt (Osmeridae) are economically important in fisheries and food webs broadly across the upper Northern Hemisphere. Four species, the surf smelt *Hypomesus pretiosus*, the Japanese surf smelt *Hypomesus japonicas*, the night smelt *Spirinchus starksi*, and the capelin *Mallotus villosus* spawn in the surf at high tides on gravel beaches, even though they are not found in the intertidal zone at any other times. Each adherent egg attaches individually to sand, gravel, pebbles, or intertidal vegetation at heights that will expose them to air at low tides.

On the Pacific coast of North America, including Puget Sound, the surf smelt *Hypomesus pretiosus* forms large spawning aggregations on falling tides after an extreme high tide (Quinn et al., 2012). The species spawns over coarse sand or pebble beaches (Sweetnam et al., 2001), generally late in the afternoon or at dusk. The habit of spawning in daylight allows this species to choose oviposition sites that are shaded and provide greater protection for the incubating eggs from excessively high temperatures or desiccation (Breder & Rosen, 1966). When shaded sites are not available, egg mortality is high (Rice, 2006). Males can be found in single-sex schools on the spawning grounds before the females arrive, and males always outnumber the females in spawning aggregations. Pairs or triplets, or groups with one female and up to five males, form in these aggregations in water about 1 m deep. These parties of two to six fish catch an incoming wave and ride it to the highest point it reaches on shore. Spawning in extremely shallow water, the fish hastily release their gametes in a flurry of activity and then return to deeper water as the wave recedes. The eggs are not buried by this activity but float momentarily before adhering to the sand or gravel substrate. The eggs, attached to the substrate, remain on the beach until hatching a few weeks later (Penttila, 2001).

This species is easily caught with dip nets from shore, and although it is small, it is numerous enough to be the basis of a fishery for human consumption and for feeding aquarium and oceanarium animals (Penttila, 2007). *H. pretiosus* is sometimes called the day smelt to differentiate it from the slightly different habits of its relative, the night smelt, described below.

The Japanese surf smelt *Hypomesus japonicus*, once considered a subspecies of *H. pretiosus*, spawns on beaches in a similar fashion and may use coarse sand as a substrate rather than gravel (Hirose & Kawaguchi, 1998). Multiple males attend each female as they gather near shore; then they spawn just before dark in very shallow water during semilunar spring tides. Adherent eggs fall to the gravel and remain on the beach during incubation, although the wave backwash may send the gravel and eggs into shallow water nearby (Hirose & Kawaguchi, 1998).

Night smelt *Spirinchus starksi* is similar in adult distribution and spawning habitats to the surf smelt *H. pretiosus*, but the reproductive season of *S. starksi* is slightly earlier in the year and slightly later at night (Sweetnam et al., 2001). They prefer gravel beaches with freshwater influx near the spawning site. Spawning takes place after dark but before midnight, on falling tides. In spawning aggregations, the male-to-female ratio is highly skewed, from 10 to 1 at the start of the season to more than 100 to 1 by the end of the season. Offshore, the ratio is closer to one-to-one (Sweetnam et al., 2001). This indicates that males may approach the beach multiple times in a given spawning run, whereas females likely spawn only once during a run. Night smelt *S. starksi* are caught by recreational anglers in much the same way as surf smelt *H. pretiosus*, for much the same uses.

Another osmerid, the capelin *Mallotus villosus*, spawns on both the Atlantic and Pacific coasts of Canada and Alaska, as well as Iceland. This economically important forage fish spawns on wave-swept gravel beaches (Arimitsu et al., 2008; Hart, 1973). *M. villosus* is among the most well-studied beach-spawning fishes, as this species is an important forage fish for larger, commercially important species such as cod, haddock, and halibut, as well as seabirds and marine mammals including large whales (Nakashima & Wheeler, 2002). A commercial fishery exists in the northern Atlantic but not the Pacific for this species. The recreational fishery for *M. villosus* is very popular in Canada, and experiencing the capelin roll is a tradition for many families in Newfoundland and Labrador (Figure 4.2). The good news of the capelin roll spreads rapidly among the neighbors, causing celebration and impromptu trips to the shore for many coastal Canadians.

M. villosus approach beaches as great shoals of fish that gradually move into shallow water, spawning aquatically at the edge of the shore. Spawning in *M. villosus* cannot be predicted by the tides; both spawning and release of hatchlings require high water on shore, but apparently the push from strong wind waves is more likely to reach these heights than the tides (Frank & Leggett, 1981). Spawning occurs over only a few days in summer, but in different months at different locations within the range. Spawning occurs earlier in Europe than in North America for *M. villosus*.

M. villosus spawn in groups of two or three, with one female and two or three males. These dyads and triplets attract additional capelin as they roll in the shallow

FIGURE 4.2 (See color insert.) Capelin *Mallotus villosus* triplet rolling on a Canadian beach. Many eggs can be seen already on the gravel. (Photo by Anna Olafsdottir.)

surf together. Males are somewhat larger than females, and during the spawning season they develop two longitudinal rows of large, pointed scales above and below the lateral line that are not present in females. Mature males also develop small bumps of raised tissue on the fins, called nuptial tubercles, during the spawning season. During spawning, two males may position themselves on either side of a female and press against her, holding her between them with the scales, as the three rush toward shore on an incoming wave (Templeman, 1948). Alternatively, one male may pair with a female and spawn as a couple, or a pair may form and then a second male may attach to the other side of the female as they ride a wave toward shore (Breder & Rosen, 1966). The fish may be very briefly emerged from water by the energetic activity, but they quickly return to the sea in a subsequent wave.

M. villosus live only a few years and spawn in the second or third year of life. Most die after spawning, but others may survive a bit longer and spawn repeatedly over several years. Beach spawning capelins may be able to spawn repeatedly while capelins that spawn subtidally die afterwards (Christiansen et al., 2008; Carscadden et al., 1989). Occasionally one sees vast numbers of capelin dead on shore along the tideline. It is unlikely that they died by being stranded in air, although they may momentarily emerge during spawning. However, if many die nearshore after a spawning run, their numerous carcasses can form a somber drift line along the high tide zone and provide a generous source of carrion for terrestrial predators including bears, eagles, and other opportunistic feeders.

Eggs of *M. villosus* are adherent and stick to each other or individually to pebbles or sand. A very shallow depression may be quickly swiped into the surface of a sandy substrate by rapid side-to-side movements of the male's body and fins, but the eggs and milt are simply released into very shallow water rather than being carefully placed within a nest (Frank & Leggett, 1981). The eggs are hardy and able to withstand long periods of emergence into air under the typically cool, damp conditions of the northern coastlines of Canada, Alaska, Japan, Russia, and Iceland.

Some Atlantic populations of *M. villosus* spawn on sandy beaches (Bonlay, 2011) or spawn subtidally at a depth of about 50 to 100 m on sandy substrates where the water is very cold. Subtidal spawning may be simultaneous with beach spawning or somewhat later in the year on the Grand Banks off Canada around Iceland. Some of these subtidal areas are ancient coastlines (Carscadden et al., 1989). It is intriguing to ponder whether beach spawning occurred on these sites previously in geologic history. Perhaps these populations continued to nest in the same areas because of site fidelity, even after they were inundated by sea level rise and a changing coast. Genetic comparisons suggest the subtidal spawners are not reproductively isolated from the beach spawners, but that the mode of reproduction on beaches or subtidally may be facultative depending on the environmental conditions (Dodson et al., 1991). It has been hypothesized that subtidal spawning occurs at times when the temperatures on beaches are too high for the eggs to survive (Nakashima & Wheeler, 2002). Capelin migrate from feeding grounds offshore and then stage near the continental shelf before moving inshore to spawn on sandy beaches when the gonads mature (Olafsdottir & Rose, 2013). Egg survival at the different types of spawning sites is not equal, however. See Chapter 7 for more details. Threats to beach-spawning fishes in the future with changing climates and sea level rise are discussed in Chapter 9.

Another fish that lives in the shallow subtidal zone but spawns intertidally is the Pacific sand lance *Ammodytes hexapterus* (Ammodytidae), found in Puget Sound and the northwest Pacific coast (Haynes et al., 2007; Robards et al., 1999). The adults spend the winter months burrowed into shallow sandy intertidal substrates, perhaps for energy conservation (Winslade, 1974). This species was not known to spawn on beaches until 1989, when large dark masses of fishes were observed near shore in Puget Sound, and subsequently tiny eggs were collected from nearby gravel beaches (Penttila, 1995a,b). Culturing the eggs revealed their true identity (Pinto, 1984). *A. hexapterus* are important forage fishes for salmon to eat, but are not themselves fished by humans (Penttila, 2007).

The spawning behavior was first observed a few years ago by amateur nature videographers near a beach where this species' eggs were previously found. At high tide, groups of *A. hexapterus* can be seen gathering in shallow water and splashing about. Pairs form and they create a shallow clearing in the gravel where eggs are deposited; these spawning pits can be seen on shore when the tide recedes (Penttila, 2007).

In Japan and China are at least 25 species of puffer fishes, but only one is known to spawn on beaches (see Chapter 1). The grass puffer *Takifugu niphobles* (Tetraodontidae) lives over shallow sandy subtidal substrates and feeds on ghost shrimp (Yamahira, 1997). Before spawning, *T. niphobles* mass near shore close to gravel and cobble beaches and can be seen swimming back and forth in the water. Spawning occurs during a semilunar rising tide, after sunset but before midnight. The closer the time of the high tide is to sunset, the more concentrated in time is the spawning activity (Yamahira, 1994). The timing of spawning shifts during the season between May and August; earlier in the season, the spawning occurs a night or two before a new or full moon, but later in the season, spawning runs occur a night or two after the full or new moon phases (Yamahira, 1997).

After sunset, male *T. niphobles* begin to chase females up onto the beach. Spawning occurs when groups of 10 to 60 fish, only one of which is female, rush toward the top of a wave and briefly strand on shore for a few seconds. This rush may occur a couple of times before mating, but the splashing activity and the release of milt into the water signal that spawning has occurred (Yamahira, 1994). Each spawning run of *T. niphobles* involves multiple group spawnings and the same males returning repeatedly to the beach (Uno, 1955; Yamahira, 1994).

Mating typically occurs at the water's edge, and the adults return to deeper water in the following wave wash. However, if tides are so high that the spawning site is completely submerged, for example if the beach is backed by a seawall or other impediment, *T. niphobles* can spawn over their chosen substrate under water as well. Eggs are broadcast over the pebbles, and the water motion causes them to settle into the crevices and pockets between rocks, providing moisture and shade. There is no parental care, and survival of eggs varies widely over the course of the spawning season (Yamahira, 1996).

The species in this group are famous for their neurotoxic tissues. In particular the ovaries and the eggs themselves are toxic. Avoidance of predators has been suggested among the many potential selective advantages to beach-spawning fishes, but for *T. niphobles*, the toxic eggs do not need to be placed on a beach to avoid being eaten by a predator. Other selection pressures, such as increased oxygen availability and warmer temperatures, may underlie the beach-spawning behavior. An aquarium that was set up with a sloping gravel bed and artificial tides was able to induce aggregating behaviors in captive *T. niphobles*, but no spawning occurred in the aquarium (Motohashi et al., 2010).

A different species, the tiger puffer *Takifugu rubripes*, is an important food species and can be cultured in captivity including spawning and rearing (Kotani et al., 2011). This species, when kept in net pens throughout life and fed a controlled diet, is not toxic like its wild relatives, indicating that the toxin is obtained extrinsically from food sources (Noguchi et al., 2006).

4.3   SOME FISHES MIGRATE ONTO SANDY BEACHES TO SPAWN

New world Silversides, Atherinopsidae, comprise small silver forage fishes that include four species that spawn on beaches. The California grunion *Leuresthes tenuis* is perhaps the most dramatic example of beach spawning in fishes, as both the spectacular spawning run and the cryptic embryonic incubation occur completely out of water (Walker, 1952). Endemic to the outer coast of California and Baja California, *L. tenuis* males and females surf up onto sandy beaches in waves (Walker, 1952). Females, using their tails to burrow vertically into the soft wet sand, emit their eggs beneath the surface (Figure 4.3). At the same time, one or more males gather around her and provide milt that flows through the watery sand to the eggs (Spratt, 1986). Individual fish are out of water for only a few minutes, but males may return repeatedly to spawn during a run (Walker, 1952).

Males generally outnumber females on the beach, at times by a factor of 8 to 10. In *L. tenuis*, with massive spawning aggregations and external fertilization, multiple paternity is common within a clutch (Byrne & Avise, 2009). In parallel, an individual male may fertilize eggs from multiple females; his genital papilla allows repeated controlled releases of sperm (Aryafar, unpublished). In these situations sperm competition is likely. This promiscuous mating provides high levels of genetic diversity in the next generation (Gaida et al., 2003; Johnson et al., 2009; Byrne et al., 2013).

The clutches of eggs of *L. tenuis* remain buried under the sand but out of water throughout incubation, the coarse substrate providing a way to keep both moisture and oxygen levels high (Martin et al., 2009). Because of the timing of the spawning runs, oviposition in the beach sand is at a tidal height that will not be submerged again for over a week, until the rising tides of the next new or full moon phase uncovers the eggs and frees them to hatch. Hatching is extremely rapid, environmentally cued by agitation in seawater (Griem & Martin, 2000; Speer Blank & Martin, 2004). If the cue fails to arrive on schedule, incubation can be extended for days or weeks (Darken et al., 1998; Martin, 1999; Moravek & Martin, 2011). See Chapter 8 for more on hatching.

Spawning runs occur late at night on falling tides after the highest tides of both new and full moons, and may last for up to an hour or more, a living river of silver fish lining the starlit shore. Being a predictable and spectacular show, grunion runs of *L. tenuis* are well publicized in local media and are the subject of educational programs (Martin et al., 2006, 2011), aquarium tours (Harzen et al., 2011), and recreational

(a)

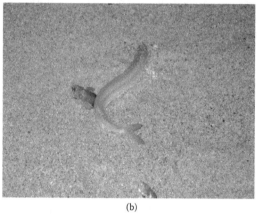

(b)

FIGURE 4.3 (See color insert.)   (a) A spawning run of California grunion *Leuresthes tenuis* out of water on a sandy beach. (Photo by Bill Hootkins.) (b) The female digs in vertically while the male fertilizes the eggs from the sand surface. (Photo by Doug Martin.)

fishing (Spratt, 1986). Out of water during the spawning runs, the *L. tenuis* are protected by a short closed season. During the open season, unique restrictions ban any kind of gear. Nothing may be used to capture the *L. tenuis* legally except bare hands, and anyone 16 years of age and older requires a fishing license. However, during open season, hundreds of anglers on a beach may outnumber the *L. tenuis* attempting to spawn, and every fish that appears may be captured. See Chapter 9 for more on threats to this species, and Chapter 10 for more about conservation efforts for this species.

The congener of *L. tenuis*, the Gulf grunion *Leuresthes sardina,* spawns at the water's edge of the Gulf of California, where wave action is less energetic (Thomson & Muench, 1976). *L. tenuis* sometimes spawn within bays that have low wave action; in these cases they behave similarly at the water's edge rather than dropping out of receding waves and fully emerging as they do on the open coast (Martin et al., 2013).

Spawning occurs after high semilunar syzygy tides, and eggs are buried in beach sand in similar fashion to *L. tenuis*. Embryos of *L. sardina* also incubate out of water until the following semilunar tides (Moffatt & Thomson, 1978). The two species may have split during a vicariant event when populations were separated by the peninsula of Baja California and changes in sea level (Bernardi et al., 2003).

Spawning runs for *L. sardina* occur on the same days and similar times of the tidal cycle as for *L. tenuis*, but because of their location at the apex of the Gulf of California, some spring high tides occur during the daytime, and therefore so do the spawning runs. For some reason these runs have not captured the imagination of the public or anglers in Baja California, but are a curiosity for tourists and any people who happen upon them. Unfortunately, the protections for *L. tenuis* in the United States do not exist for either of the *Leuresthes* species in Mexico.

4.4   SOME FISHES SPAWN ON INTERTIDAL OR NEARSHORE VEGETATION

A third beach-spawning silverside, *Menidia menidia*, the Atlantic silverside, spawns over sea grass beds at high tide along the Atlantic coast of the United States (Middaugh, 1981). Large aggregations form near shore, attracting avian and terrestrial predators. After spawning occurs, the eggs adhere to the vegetation and are tidally emerged during low tides (Middaugh et al., 1983).

The presence of so many *M. menidia* spawning turns the water white with milt, and the high activity depletes the oxygen dissolved in the water, causing the fish to act as if they are in a "spawning stupor" that leaves them vulnerable to predators (Middaugh, 1981). See Chapter 5 for more on this.

The presence of appropriate substrate for oviposition is vital for spawning to occur in *M. menidia*. Because spawning is aquatic, the substrate must be submerged underwater, so on most of the Atlantic coast, the timing of reproduction is associated with high tides. However, in the laboratory, with food *ad libitum* and an artificial substrate always available, spawning can occur much more frequently (Conover & Kynard, 1984).

M. menidia has environmental sex determination (Conover & Kynard, 1981), with temperature the important factor in deciding whether an individual will become male or female. As with many beach-spawning fishes, males outnumber females on the spawning grounds. In the laboratory, over several generations of selection, the sex ratio of a given temperature can evolve to maintain adequate numbers of both sexes for successful reproduction (Conover et al., 1992). This plasticity may help this species adjust to climate change.

The fourth species of Atherinopsidae that spawns on beaches is the false grunion, *Colpichthys regis*, an estuarine fish from the Gulf of California (Russell et al., 1987). This species has been observed high in the intertidal zone at high semilunar tides, spawning over vegetation or among some tiles that were left as construction rubble. The demersal eggs fell or rolled into cracks and crevices, finding shade and protection from desiccation. The embryos incubate out of water in these sheltered microhabitats and hatch when submerged at a subsequent high tide. This species spawns in the morning in winter months. The spawning season is probably limited by temperature tolerances of the embryos.

The presence of four species that spawn on beaches within the same family is impressive, but the three genera are distantly related, and each probably represents an independent origin of beach-spawning behavior (Martin & Swiderski, 2001). See Chapter 1 for more on the evolution of beach spawning.

The Pacific herring *Clupea pallasi pallasi* (Clupeidae), an economically important forage fish, spawns intertidally (Jones, 1972). *C. p. pallasi* release their adherent eggs over sea grass beds (*Zostera* spp.) that are intertidal or in shallow subtidal waters (Lassuy & Moran, 1989), or over worm tubes or other complex substrates that will not bury the eggs under sediment. Vast shoals of herring attract commercial fishermen, seabirds, and shoreline observers in San Francisco Bay and other spawning sites every winter (Haegele & Schweigert, 1985). Spawning takes place aquatically, in a high-energy wave environment near shore.

Females contact the substrate to oviposit. In some populations of Atlantic herring, the males also contact the substrate (Haegele & Schweigert, 1985). The embryos in eggs are exposed to air tidally in the low intertidal zone, and survival is greater for eggs that are intertidal than for those that are subtidal (Stacey & Hourston, 1982). The survival of eggs and embryos depends on avoiding suffocation from overcrowding of the eggs, or from silting in of the oviposition site. Exposure to air for too long can also destroy the embryos.

Pacific herring spawn near shore in areas from San Francisco Bay north to Alaska, and across the Pacific in Japan and the Sea of Okhotsk (Haegele & Schweigert, 1985). The spawn is collected from seaweed and sea grass during low tides for food by native Alaskans and is served as sushi in Japan. The massive Exxon Valdez oil spill profoundly affected the herring spawn in the intertidal areas of Alaska's Prince William Sound in 1989, as did the Cosco Busan fuel oil spill on San Francisco Bay in 2007 (Incardona et al., 2011). See Chapter 9 for more on threats and challenges to beach-spawning fishes.

4.5 SOME BEACH-SPAWNING FISHES ARE ANADROMOUS OR CATADROMOUS

Most beach-spawning fishes live near the ocean shore, in either the intertidal or subtidal zones. However, a few species make long migrations from other habitats to spawn on beaches.

Catadromous fishes live in freshwater as adults but migrate into estuaries or seawater to spawn. The whitebait *Galaxias maculatus* (Galaxiidae) of New Zealand migrates downstream in massive numbers to spawn in the high intertidal zone of estuaries, on fringing vegetation (McDowall, 1968, 1991). Eggs are tidally emerged and hatch on immersion at the following semilunar tides. Most adults of this salmoniform group spawn once, apparently cued by the lunar cycle and the tides, and then die, but some may live on and return upriver. The larvae spend several months at sea before swimming into freshwater to mature.

Many other species' eggs may be out of water for some time during development. A congener, the landlocked, nonmigratory *Galaxias fasciatus*, stays in freshwater streams throughout life and does not synchronize its reproduction to the tides or moon cycles (Hopkins, 1979). *Galaxias attenuatus* and *Enchelyopus cimbrus*

(Lotidae) also may incubate eggs partially out of water (Yamagami, 1988; Munro et al., 1990).

Anadromous fishes hatch out in streams and then migrate and grow to maturity in the ocean, usually over several years. Then they return into freshwater streams to spawn. Atlantic and Pacific salmon are the classic examples of anadromous fishes. One species in Prince William Sound and southeast Alaska, the chum salmon *Oncorhynchus keta* (Salmonidae), spawns in the intertidal zone of the mouths of streams in areas where freshwater upwells (NOAA Fisheries Report, 2005). The spawning pairs build redds and lay eggs during lower tides, when freshwater prevails, from late June to September (NOAA Fisheries Report, 2005). Both parents die after spawning. The freshwater upwelling around the eggs prevents the embryos from freezing during the incubation period, throughout winter and into the following May.

With this form of beach spawning, the *O. keta* eggs are not emerged into air by tides. Oviposition takes advantage of the location of freshwater springs at an appropriate tidal height on beaches. This behavior drastically shortens the spawning migration for the adults and the downstream migration of the juveniles, while still providing the necessary clear, flowing freshwater current in which their embryos can develop. (Clever fish!) The larvae spend several months in the adjacent estuary before moving out to the open ocean to grow and mature (NOAA Report, 2005).

4.6 SOME FISHES APPEAR TO BE TRANSITIONING TOWARD BEACH SPAWNING

As we have seen, beach-spawning behavior is widespread among fish families but present in only one or a few species within each group (Martin & Swiderski, 2001). Behavioral plasticity should allow rapid evolution of novel spawning behaviors and use of nontraditional oviposition sites. See Chapter 1 for more on the evolution of beach spawning. Following are several examples of evolution apparently in progress.

Stickleback fishes (Gasterosteidae) are model organisms for the study of courtship and parental care in fishes. Many sticklebacks are anadromous; others live fully in freshwater or in the shallow marine subtidal zone. The *Gasteroteus* species complex of small, darkly colored fishes has been well-studied for reproductive behavior. Males provide attentive courtship and parental care. A typical stickleback male builds an elaborate nest on rocks or pebbles on the floor of the water body over the course of several days. He then displays energetic courtship behaviors to attract a gravid female. After she lays eggs within the nest, the male follows and fertilizes the eggs, and then he cares for them up to and after hatching (Ostlund-Nilsson et al., 2007). While caring for the eggs, the father chases away other males but may mate with another female, adding her eggs to the same nest.

One recently discovered population, called the white stickleback for the conspicuous bright white male coloration during the breeding season, is related to *Gasterosteus aculeatus* and lives sympatrically on the Canadian coast (MacDonald et al., 1995a). Males, in addition to the startling color, differ from the other populations by not building nests and providing no additional parental care beyond scattering the fertilized eggs (Blouw, 1996). More typical *G. aculeatus* build nests on the substrate.

White sticklebacks were first observed spawning in the shallow subtidal zone on filamentous algae, instead (Blouw & Hagen, 1990). The males sometimes moved the spawned eggs and dispersed them around the subtidal algae filaments but did not build nests (MacDonald et al., 1995b).

Within the population of white sticklebacks, a subgroup in Nova Scotia was observed spawning in the intertidal zone in the early 1990s (MacDonald et al., 1995a). These white sticklebacks spawned over bare rock in the rocky intertidal zone, a habitat that was not previously known to have a nursery function for this genus (MacDonald et al., 1995a,b). Males either built no nests or made flimsy constructs from available materials, and then dispersed the nonadherent eggs into crevices in the bare gravel substrate. Embryos were able to develop within the eggs and hatch under these conditions on the beach, submerged and emerged by tides. However, they were vulnerable to high temperatures and desiccation (MacDonald et al., 1995b).

The breeding behavior was observed in the intertidal zone over several years in this location (MacDonald et al., 1995a). Individual white stickleback males appeared to prefer one spawning substrate or the other, either the subtidal algae or the intertidal bare rock, but did not use both. The preference for unvegetated rock substrate was sustained in the laboratory, even when ample filamentous algae was available in the aquarium. In the laboratory, males that preferred spawning over algae excluded the males that were seen spawning over bare rock. Females were attracted to both types of males and provided eggs for both substrates.

This intriguing situation shows three different types of spawning behavior and oviposition substrates within one population. These changes in behavior may result in reproductive isolation even before genetic divergence has occurred. Although the white stickleback has not been distinguished as a separate species from sympatric G. aculeatus with allozyme data (Haglund et al., 1990), it is reproductively isolated and is currently under study for the possibility that it may be an incipient species.

Another example of a species that may be transitioning into intertidal spawning is the righteye flounder, Lepidopsetta bilineata (Pleuronectidae), sometimes called the rock sole (Penttila, 1995b). Off the coast of Alaska is one of the largest flatfish fisheries in the United States. Females spawn over a variety of different substrates including sand, mud, and intertidal shores. The eggs stick where they land and may hatch within a week to 25 days, depending on temperatures. In most areas, spawning is subtidal; however, in some areas of the Salish Sea south of Seattle, L. bilineata spawns at high tides at the upper and mid beach (D. Penttila, personal communication).

In winter, on some beaches the flatfish eggs may outnumber those of sympatric beach-spawning fishes, the Pacific sand lance Ammodytes hexapterus and the surf smelt Hypomesus pretiosus. L. bilineata is among the most common nearshore flatfishes in Puget Sound, but intertidal spawning behavior has not yet been observed. After the eggs had been collected repeatedly from gravel beaches, the identity of this previously unknown type of egg was discovered by rearing the eggs and larvae in the Seattle Aquarium until positive identification could be made (D. Penttila, personal communication). This is the only known flatfish that spawns on beaches. Because flatfish frequently visit beaches during high tides to feed on invertebrates, it is likely that other species of beach-spawning flatfishes are yet to be discovered.

A likely possibility for another twist on beach spawning is found in a couple of well-studied gobies (Gobiidae), the blueband goby *Lythrypnus dalli* and the striped goby *L. zebra*. Found on rocky reefs both intertidally and subtidally, *L. dalli* has an intriguing form of bidirectional sex change that also occurs in *L. zebra* in captive individuals (St. Mary, 1993). For small gobies that are vulnerable to predation, taking on the appropriate gender role with a new mate in an available territory may be more conducive to survival than maintaining the same gender through time (Munday et al., 2010).

Among the *Lythrypnus* species, individuals may be simultaneous hermaphrodites, holding both ovarian and testicular tissues in the same individual, although each individual functions as only one gender at a time over repeated breeding encounters (St. Mary, 1998). The ovarian tissues increase during a change to female and decrease while testicular tissues increase during a change to male. *L. dalli* can reverse gender and change in either direction repeatedly, depending on whether the individual is dominant or subordinate in a given pair (Rodgers et al., 2007). *L. zebra* has similar tissue characteristics to *L. dalli* (St. Mary, 1993). Both *L. dalli* and *L. zebra* nest in cavities and seem likely candidates for intertidal spawning in their warm coastal habitats along California and Mexico. If so, this would be the first known case of hermaphroditism in intertidal or beach-spawning fishes, although many variations on sequential and simultaneous hermaphroditism are known from numerous species of fishes, including gobies that spawn on coral reefs and other subtidal substrates (Warner, 1988).

4.7 SUMMARY FOR FISHES THAT MIGRATE FROM OTHER HABITATS TO SPAWN ON BEACHES

Species of fishes from many different lineages make spawning migrations onto beaches from habitats that include the deeper ocean, freshwater, and estuarine waters. In general large aggregations of the species form near the beach site shortly before spawning occurs. Spawning takes place on rocky beaches, on gravel beaches, on sandy beaches, over intertidal vegetation and other biogenic habitat, or onto bare substrates (Figure 4.1). The appropriate substrate at the appropriate tidal height may be accessible to the fishes only during high tides. Typically the fishes spawn in shallow water, but not always; for example, *Leuresthes tenuis* spawns terrestrially on wave-swept beaches. Eggs may be carefully attached individually in a monolayer, buried under a thin layer of sand or gravel, or scattered and stuck onto nearby objects.

Although pairs form in most spawning aggregations, the pair may be joined by additional males during the spawning rush to shore. Group spawning is present in many of these species. Sperm competition is likely, and multiple paternity of clutches may be common. The males on shore vastly outnumber the females during a spawning run, as the females in general spawn once while the males repeatedly approach the beach during a given run.

Beach-spawning fishes place their eggs on characteristic substrates at specific tidal heights on shore. To reach these heights typically requires a semilunar syzygy high tide, except in the case of *Mallotus villosus,* which depends on wind waves. The timing of capelin runs is not as predictable as that of runs synchronized by the tides (Martin et al., 2004; Figure 4.4). Adherent eggs may be hidden under boulders or buried in sand, or left to roll free without additional protection.

Parental care beyond oviposition does not occur in these fishes except for those few that migrate into the rocky intertidal zone, and in those cases the amount of protection given the eggs is variable between species. Parental care is solely a male prerogative in these migrant fishes, and nest guarding seems to aid his reproductive fitness by providing opportunities for him to mate with additional females.

The long migrations, very precise placement of eggs, and variety of species that find their way to beaches from other habitats all point to a conclusion that the advantages outweigh the disadvantages, and the benefits balance the costs for this intriguing reproductive behavior. However, there are still many risks involved. See Chapter 6 and Chapter 9 for more about the kinds of trouble waiting for these fishes at their spawning grounds.

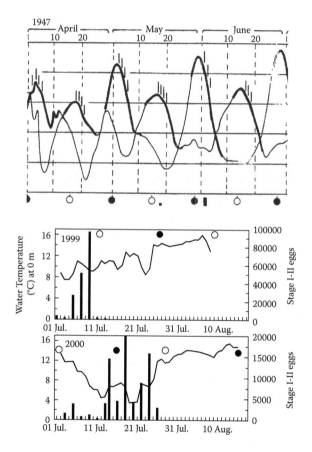

FIGURE 4.4 Comparison of the timing of spawning runs for tidally synchronized California grunion, *Leuresthes tenuis* (top panel), and capelin *Mallotus villosus* that depend on wind waves instead (middle and lower panels). (After Martin et al. (2004), from Walker (1949) and Nakashima & Wheeler (2002)).

REFERENCES

Arimitsu, M. L., Piatt, J. F., Litzow, M. A., Abookire, A. A., Romero, M. D., & Robards, M. D. (2008). Distribution and spawning dynamics of capelin (*mallotus villosus*) in Glacier Bay, Alaska: A cold water refugium. *Fisheries Oceanography* 17, 137–146.

Arora, H. L. (1948). Observations on the habits and early life history of the Batrachoidid fish, *Porichthys notatus* Girard. *Copeia* 1948, 89–93.

Aryafar, H. (2012). *Sexual Dimorphism in Reproductive Morphology of Leuresthes tenuis (Atherinopsidae)*. Master's thesis. California State University, Fullerton.

Bailey, J. E., Wing, B. G. & Landingham, J. H. (1983). Juvenile Pacific sandfish, *Trichodon trichodon*, associate with pink salmon, *Oncorhynchus gorbuscha*, fry in the nearshore area, southeastern Alaska. *Copeia* 1983, 549–551.

Bass, A. H. & Marchaterre, M. A. (1989). Sound-generating (sonic) motor system in a teleost fish (*Porichthys notatus*): Sexual polymorphism in the ultrastructure of myofibrils. *Journal of Comparative Neurology* 286, 141–153. DOI: 10.1002/cne.902860202.

Bernardi, G., Findley, L. & Rocha-Olivares, A. (2003). Vicariance and dispersal across Baja California in disjunct marine fish populations. *Evolution* 57, 1599–1609.

Blouw, D. M. (1996). Evolution of offspring desertion in a stickleback fish. *Ecoscience* 3, 18–24.

Blouw, D. M. & Hagen, D. W. (1990). Breeding ecology and evidence of reproductive isolation of a widespread stickleback fish (Gasterosteidae) in Nova Scotia, Canada. *Biological Journal of the Linnean Society* 39, 195–217. DOI: 10.1111/j.`095-8312.1990.tb00512.x.

Boulay, C. (2011). Capelin Observers Network Observer Kit 2012. Fisheries and Oceans Canada. http://www.qc.dfo-mpo.gc.ca/signaler-report/roc-con/2011-2012/Observer_kit.2012.pdf.

Brantley, R. K. & Bass, A. H. (1994). Alternative male spawning tactics and acoustic signaling in the plainfin midshipman fish, *Porichthys notatus*. *Ethology* 96, 213–232.

Breder, C. M. & Rosen, D. E. (1966). *Modes of Reproduction in Fishes*. Garden City: Natural History Press.

Byrne, R. & Avise, J. (2009). Multiple paternity and extra-group fertilizations in a natural population of California grunion (*Leuresthes tenuis*), a beach-spawning marine fish. *Marine Biology* 156, 1681–1690.

Byrne, R., Bernardi, G. & Avise, J. C. (2013). Spatiotemporal genetic structure in a protected marine fish, the California grunion (*Leuresthes tenuis*), and relatedness in the genus *Leuresthes*. *Journal of Heredity* 104, 521–531. DOI: 10.1093/jhered/est024.

Carscadden, J. E., Frank, K. T. & Miller, D. S. (1989). Capelin (*Mallotus villosus*) spawning on the southeast shoal: influence of physical factors past and present. *Canadian Journal of Fisheries and Aquatic Sciences* 46, 1743–1754.

Christiansen, J., Praebel, K., Siikavuopio, S. I., & Carscadden, J. E. (2008). Facultative semelparity in capelin *Mallotus villosus* (Osmeridae)—an experimental test of a life history phenomenon in a subarctic fish. *Journal of Experimental Marine Biology and Ecology* 360, 47–55. DOI: 10.1016/j.jembe.2008.04.003.

Coleman, R. M. (1999). Parental care in intertidal fishes. In *Intertidal Fishes: Life in Two Worlds*, (Horn, M. H., Martin, K. L. M. & Chotkowski, M. A., eds.), 165–180. San Diego: Academic Press.

Conover, D. O. & Kynard, B. E. (1981). Environmental sex determination: interaction of temperature and genotype in a fish. *Science* 213, 577–579.

Conover, D. O. & Kynard, B. E. (1984). Field and laboratory observations of spawning periodicity and behavior of a northern population of the Atlantic silverside, *Menidia menidia* (Pisces: Atherinidae). *Environmental Biology of Fishes* 11, 161–171.

Conover, D. O., Van Voorhees, D. A. & Ehtisham, A. (1992). Sex ratio selection and the evolution of environmental sex determination in laboratory populations of *Menidia menidia*. *Evolution* 46, 1722–1730.

Crane, J. M., Jr. (1981). Feeding and growth by the sessile larvae of the teleost *Porichthys notatus*. *Copeia* 1981, 895–897.

Darken, R. S., Martin, K. L. M. & Fisher, M. C. (1998). Metabolism during delayed hatching in terrestrial eggs of a marine fish, the grunion *Leuresthes tenius*. *Physiological Zoology* 71, 400–406.

DeMartini, E. E. (1978). Spatial aspects of reproduction in buffalo sculpin, *E. bison*. *Environmental Biology of Fishes* 3:331–336.

DeMartini, E. E. (1988). Spawning success of the plainfin midshipman. I. Influences of male body size and area of spawning site. *Journal of Marine Biology and Ecology* 121, 177–192.

DeMartini, E. E. (1999). Intertidal spawning. In *Intertidal Fishes:Life in Two Worlds*, (Horn, M. H., Martin, K. L. M. & Chotkowski, M. A., eds.), 143–164. San Diego: Academic Press.

Dodson, J. J., Carscadden, J. E., Bernatchez, L. & Colombani, F. (1991). Relationship between spawning mode and phylogeographic structure in mitochondrial DNA of North Atlantic capelin *Mallotus villosus*. *Marine Ecology Progress Series* 76: 103–113.

Eschmeyer, W. N., Herald, E. S. & Hammann, H. (1983). *A Field Guide to Pacific Coast Fishes of North America*. Peterson Field Guide Series. Boston: Houghton Mifflin.

Feder, H. M., Turner, C. H. & Limbaugh, C. (1974). *Observations on fishes associated with kelp beds in southern California. Fish Bulletin* 160. State of California Resources Agency, Department of Fish & Game. http://content.cdlib.org/view?docId=kt9t1nb3s h&brand=oac4.

Fine, M. L., Malloy, K. L., King, C. B., Mitchell, S. L. & Cameron, T. M. (2001). Movement and sound generation by the toadfish swim bladder. *Journal of Comparative Physiology A*187, 371–379.

Frank, K. T. & Leggett, W. C. (1981). Wind regulation of emergence times and early larval survival in capelin (*Mallotus villosus*). *Canadian Journal of Fisheries and Aquatic Sciences* 38, 215–223.

Gaida, I., Buth, D. G., Matthews, S. D., Snow, A. L., Luo, S. B. & Kutsuna, S. (2003). Allozymic variation and population structure of the California grunion, *Leuresthes tenuis* (Atheriniformes: Atherinopsidae). *Copeia* 2003, 594–600.

Gibson, R. N. (1982). Recent studies on the biology of intertidal fishes. *Oceanography and Marine Biology Annual Reviews* 20, 363–414.

Griem, J. N. & Martin, K. L. M. (2000). Wave action: the environmental trigger for hatching in the California grunion *Leuresthes tenius* (Teleostei: Atherinopsidae). *Marine Biology* 137, 177–181.

Haegele, C. W. & Schweigert, J. F. (1985). Distribution and characteristics of herring spawning grounds and description of spawning behavior. *Canadian Journal of Fisheries and Aquatic Sciences* 42(S1): s39–s55, 10.1139/f85–261.

Haglund, T. R., Buth, D. G. & Blouw, D. M. (1990). Allozyme variation and the recognition of the "white stickleback." *Biochemical Systematics and Ecology* 18, 559–563. http://dx.doi.org/10.1016/0305-978(90)90129-4.

Hart, J. L. (1973). *Fishes of the Pacific coast of Canada*. Fisheries Research Board of Canada Bulletin 180, 1–740.

Harzen, S. E., Brunnick, B. J. & Schaadt, M. (2011). *An Ocean of Inspiration: The John Olguin Story*. Toronto, Canada: Rocky Mountain Books, 324 pp.

Haynes, T. B., Ronconi, R. A., & Burger, A. E. (2007). Habitat use and behavior of the Pacific Sandlance (*Ammodytes hexapterus*) in the shallow subtidal region of southwest Vancouver Island. *Northwestern Naturalist* 88, 155–167.

Hirose, T. & Kawaguchi, K. (1998). Spawning ecology of Japanese surf smelt, *Hypomesus pretiosus japonicas* (Osmeridae), in Otsuchi Bay, northeastern Japan: *Environmental Biology of Fishes* 52, 213–223.

Hopkins, C. L. (1979). Reproduction of *Galaxias fasciatus* Gray (Salmoniformes: Galaxiidae), *New Zealand Journal of Marine and Freshwater Research* 13, 225–230.

Incardona, J. P., Vines, C. A., Anulacion, B. F., Baldwin, D. H., Day, H. L., French, B. L., Labenia, J. S., et al. (2011). Unexpectedly high mortality in Pacific herring embryos exposed to the 2007 *Cosco Busan* oil spill in San Francisco Bay. *Proceedings of the National Academy of Sciences Plus* 109, E51–E58. doi/10.1073/pnas.1108884109.

Johnson, P. B., Martin, K. L., Vandergon, T. L., Honeycutt, R. L., Burton, R. S. & Fry, A. (2009). Microsatellite and mitochondrial genetic comparisons between northern and southern populations of California grunion *Leuresthes tenuis*. *Copeia* 2009, 467–476.

Jones, B. C. (1972). Effect of intertidal exposure on survival and embryonic development of Pacific herring spawn. *Journal of the Fisheries Research Board of Canada* 29, 1119–1124.

Kent, D. I., Fisher, J. D. & Marliave, J. B. (2011). Interspecific nesting in marine fishes: spawning of the spinynose sculpin, *Asemichthys taylori*, on the eggs of the buffalo sculpin, *Enophrys bison*. *Ichthyological Research* 58, 355–359.

Knapp, R., Wingfield, J. C. & Bass, A. H. (1999). Steroid hormones and paternal care in the plainfin midshipman fish (*Porichthys notatus*). *Hormones and Behavior* 35, 81–89.

Kotani, T., Wakiyama, Y., Imoto, T. & Fushimi, H. (2011). Improved larviculture of ocellate puffer *Takifugu rubripes* through control of stocking density. *Aquaculture* 213, 95–101.

Lassuy, D. R. & Moran, D. (1989). Species profiles: life histories and environmental requirements of coastal fishes and invertebrates (Pacific Northwest)—Pacific herring. *U. S. Fish and Wildlife Service Biological Reports* (11.126) U.S. Army Corps of Engineers, TR-EL-82-4.

Love, M. S. (2011). *Certainly More Than You Want to Know about the Fishes of the Pacific Coast: A Postmodern Experience.* Santa Barbara, CA: Really Big Press, 672 pp.

MacDonald, J. F., Bekkers, J., MacIsaac, S. M. & Blouw, D. M. (1995a). Intertidal breeding and aerial development of embryos of a stickleback fish (*Gasterosteus*). *Behaviour* 132, 1183–1206. http://www.jstor.org/stable/453531.

MacDonald, J. F., MacIsaac, S. M., Bekkers, J. & Blouw, D. M. (1995b). Experiments on embryo survivorship, habitat selection, and competitive ability of a stickleback fish (*Gasterosteus*) which nests in the rocky intertidal zone. *Behaviour* 132, 1207–1221. http://www.jstor.org/stable/453533.

Marliave, J. B. (1980). Spawn and larvae of the Pacific sandfish, *Trichodon trichodon*. *Fisheries Bulletin US* 78, 959–964.

Marliave, J. B. (1981). High intertidal spawning under rockweed, *Fucus distichus*, by the sharpnose sculpin, *Clinocottus acuticeps*. *Canadian Journal of Zoology* 59, 1122–1125.

Martin, K., Speer-Blank, T., Pommerening, R., Flannery, J. & Carpenter, K. (2006). Does beach grooming harm grunion eggs? *Shore & Beach* 74, 17–22.

Martin, K. L. M. (1999). Ready and waiting: delayed hatching and extended incubation of anamniotic vertebrate terrestrial eggs. *American Zoologist* 39, 279–288.

Martin, K. L. M. & Swiderski, D. L. (2001). Beach spawning in fishes: Phylogenetic tests of hypotheses. *American Zoologist* 41, 526–537.

Martin, K. L. M., Bailey, K., Moravek, C. & Carlson, K. (2011). Taking the plunge: California grunion embryos emerge rapidly with environmentally cued hatching. *Integrative and Comparative Biology* 51, 26–37. DOI: 10.1093/icb/icr037.

Martin, K. L. M., Heib, K. A. & Roberts, D. A. (2013). A southern California icon surfs north: local ecotype of California grunion, *Leuresthes tenuis* (Atherinopsidae) revealed by multiple approaches during temporary habitat expansion into San Francisco Bay. *Copeia* 2013, 729–739. DOI: 10.1643/CI-13-036.

Martin, K. L., Moravek, C. L. & Walker, A. J. (2010). Waiting for a sign: extended incubation postpones larval stage in the beach spawning California grunion *Leuresthes tenuis* (Ayres). *Environmental Biology of Fishes* 91, 63–70. doi./10.1007/s10641-010-9760-4.

Martin, K. L. M., Van Winkle, R. C., Drais, J. E. & Lakisic, H. (2004). Beach spawning fishes, terrestrial eggs, and air breathing. *Physiological & Biochemical Zoology* 77, 750–759.

McDowall, R. M. (1968). *Galaxias maculatus* (Jenyns), the New Zealand whitebait. *New Zealand Marine Department of Fisheries Research Bulletin* 2, 1–84.

McDowall, R. M. (1991). *Conservation and Management of the Whitebait Fishery*. Science and Research Series No 38. New Zealand: Department of Conservation, 18 pp.

Middaugh, D. P. (1981). Reproductive ecology and spawning periodicity of the Atlantic silverside, *Menidia menidia* (Pisces: Atherinidae). *Copeia* 1981, 766–776.

Middaugh, D. P., Kohl, H. W. & Burnett, L. E. (1983). Concurrent measurement of intertidal environmental variables and embryo survival for the California grunion, *Leuresthes tenuis*, and Atlantic silverside, *Menidia menidia* (Pisces: Atherinidae). *California Fish & Game* 69, 89–96.

Moffatt, N. M. & Thomson, D. A. (1978). Tidal influence on the evolution of egg size in the grunions (*Leuresthes*, Atherinidae). *Environmental Biology of Fishes* 3, 267–273.

Moravek, C. L. & Martin, K. L. (2011). Life goes on: delayed hatching, extended incubation and heterokairy in development of embryonic California grunion *Leuresthes tenuis*. *Copeia* 2011, 308–314. DOI: 10.1643/CG-10-164.

Motohashi, E., Yoshihara, T., Doi, H. & Ando, H. (2010). Aggregating behavior of the grass puffer, *Takifugu niphobles*, observed in aquarium during the spawning period. *Zoological Sciences* 27, 559–564. DOI: 10.2108/zsj.17.559.

Munday, P. L., Kuwamura, T. & Kroon, F. J. (2010). Bidirectional sex change in marine fishes. In *Reproduction and Sexuality in Marine Fishes* (Cole, K.S., ed.), 241–271. Berkeley: University of California Press.

Munro, A. D., Scott, A. P. & Lam, T. J. (Eds). (1990). *Reproductive Seasonality in Teleosts: Environmental Influences*. Boca Raton: CRC Press.

Nakashima, B. & Wheeler, J. P. (2002). Capelin (*Mallotus villosus*) spawning behavior in Newfoundland waters— the interaction between beach and demersal spawning. *ICES Journal of Marine Sciences* 59, 909–916. DOI: 10.1006/jmsc.2002.1261.

NOAA Fisheries Report (2005). *Environmental Impact Statement for Essential Fish Habitat Identification and Conservation in Alaska*. www.fakr.noaa.gov/habitat/seis/final/Volume_I/Chapter_3.pdf.

Noguchi, T., Arakawa, O. & Takatani, T. (2006). Toxicity of pufferfish *Takifugu rubripes* cultured in net-cages at sea or aquaria on land. *Comparative Biochemistry and Physiology D Genomics and Proteomics* 2006, 1, 153–157. DOI: 10.1016/j.cbd.2005.11.003.

Olafsdottir, A. H. & Rose, G. A. (2013). Staged spawning migration in Icelandic capelin (*Mallotus villosus*): effects of temperature, stock size and maturity. *Fisheries Oceanography* 22, 446–458. DOI: 10.1111/fog.12032.

Ostlund-Nilsson, S., Mayer, I. & Hunting Ford, F. A. (eds.). (2007). *Biology of the three-spined stickleback*. Boca Raton: CRC Press, 408 pp.

Penttila, D. E. (1995a). Investigations of the spawning habitat of the Pacific sand lance, *Ammodytes hexapterus*. In *Puget Sound Research '95 Conference Proceedings*, 855–859. Olympia, WA: Puget Sound Water Quality Authority.

Penttila, D. E. (1995b). The WDFW's Intertidal Baitfish Spawning Beach Survey Project in Puget Sound. In *Proceedings of the Puget Sound Research '95 Conference*, 235–241. Puget Sound Water Quality Authority, Olympia, WA, Vol. 1.

Penttila, D. E. (2001). Intertidal spawning ecology of three species of marine forage fishes in Washington State. *Journal of Shellfish Research* 20, 1198.

Penttila, D. E. (2007). Marine forage fishes in Puget Sound. Technical Report 2007–03, Washington Department of Fish and Wildlife, Washington Sea Grant, 23 pp. Olympia, WA: Puget Sound Nearshore Partnership.

Petersen, C. W. & Hess, H. C. (2011). Evolution of parental behavior, egg size, and egg mass structure in sculpins. In *Adaptation and Evolution in Cottoid Fishes* (Goto, A., Munehara, H. & Yabe, M., eds.), 194–203. Tokai University Press, Kanagawa, Japan.

Petersen, C. W., Salinas, S., Preston, R. L. & Kidder, G. W. III. (2010). Spawning periodicity and reproductive behavior of *Fundulus heteroclitus* in a New England salt marsh. *Copeia* 2010, 203–210. DOI: http://dx.doi.org/10.1643/CP-08-229.

Pinto, J. M. (1984). Laboratory spawning of *Ammodytes hexapterus* from the Pacific coast of North America with a description of its eggs and early larvae. *Copeia* 1984, 242–244.

Quinn, T., Krueger, K., Pierce, K., Penttila, D., Perry, K., Hicks, T. & Lowry, D. (2012). Patterns of surf smelt, *Hypomesus pretiosus*, intertidal spawning habitat use in Puget Sound, Washington State. *Estuaries and Coasts*, 35, 1214–1228. DOI: 10.1007/s12237-012-9511-1.

Rice, C. A. (2006). Effects of shoreline modification on a northern Puget Sound beach: microclimate and embryo mortality in surf smelt (*Hypomesus pretiosus*). *Estuaries and Coasts* 29, 63–71.

Robards, M. D., Piatt, J. F. & Rose, G. A. (1999). Maturation, fecundity, and intertidal spawning of Pacific sand lance in the northern Gulf of Alaska. *Journal of Fish Biology* 54, 1050–1068. DOI: 10.1111/j.1095-8649.1999.tb00857.x.

Rodgers, E. W., Earley, R. I. & Grober, M. S. (2007). Social status determines sexual phenotype in the bi-directional sex changing blueband goby *Lythrypnus dalli*. *Journal of Fish Biology* 70, 1660–1668.

Russell, G. A., Middaugh, D. P. & Hemmer, M. J. (1987). Reproductive rhythmicity of the Atherinid fish, *Colpichthys regis*, from Estero Del Soldado, Sonora, Mexico. *California Fish and Game* 73, 169–174.

Speer Blank, T. M. & Martin, K. L. M. (2004). Hatching events in the California grunion, *Leuresthes tenuis*. *Copeia* 2004, 21–27.

Spratt, J. D. (1986). The amazing grunion. *Marine Resource Leaflet* 3. Sacramento: California Department of Fish and Game.

Stacey, N. E. & Hourston, A. S. (1982). Spawning and feeding behavior of captive Pacific Herring, *Clupea harengus pallasi*. *Canadian Journal of Fisheries and Aquatic Sciences* 39, 489–498. DOI: 10.1139/f82-067.

St. Mary, C. M. (1993). Novel sexual patterns in two simultaneously hermaphroditic gobies, *Lythrypnus dalli* and *L. zebra*. *Copeia* 1993, 1062–1072.

St. Mary, C. M. (1998). Characteristic gonad structure in the gobiid genus *Lythrypnus* with comparisons to other hermaphroditic gobies. *Copeia* 1998, 720–724.

Sweetnam, D. A., Baxter, R. D. & Moyle, P. B. (2001). True smelts. In *California's Living Marine Resources: A Status Report*, 470–478. Sacramento: California Department of Fish and Game.

Taylor, M. (1984). Lunar synchronization of fish reproduction. *Transactions of the American Fisheries Society* 113, 484–493.

Templeman, W. (1948). The life history of the capelin (*Mallotus villosus*) in Newfoundland waters. *Bulletin of the Newfoundland Government Laboratory at St. John's* 17, 1–155.

Thedinga, J. F., Johnson, S. W. & Mortensen, D. G. (2006). Habitat, age, and diet of a forage fish in southeastern Alaska: Pacific sandfish (*Trichodon trichodon*). *Fisheries Bulletin* 104, 631–637.

Thomson, D. A. & Muench, K. A. (1976). Influence of tides and waves on the spawning behavior of the Gulf of California grunion, *Leuresthes sardina* (Jenkins and Evermann). *Bulletin of the Southern California Academy of Science* 75, 198–203.

Uno, Y. (1955). Spawning habit and early development of a puffer, *Fugu (Torafugu) niphobles* (Jordan et Snyder). Journal of Tokyo University Fisheries 42, 169–183.

Walker, B. W. (1949). *Periodicity of spawning by the grunion, Leuresthes tenuis, an atherine fish*. PhD thesis. 165 pp. University of California, San Diego: Scripps Institution of Oceanography. Retrieved from https://escholarship.org/uc/item/1j33928x.

Walker, B. W. (1952). A guide to the grunion. *California Fish & Game* 38, 409–420.

Warner, R. R. (1988). Sex change and the size-advantage model. *Trends in Ecology and Evolution* 3, 133–136.

Winslade, P. (1974). Behavioral studies on the lesser sandeel *Ammodytes marinus* (Raitt) I. The effect of food availability on activity and the role of olfaction in food detection. *Journal of Fisheries Biology* 6, 565–599.

Yamagami, K. (1988). Mechanisms of hatching in fish. In *Fish Physiology, Volume* 11A (Hoar, W. S. & Randall, D. J., eds.), 447–499. San Diego, CA: Academic Press.

Yamahira, K. (1994). Combined effects of tidal and diurnal cycles on spawning of the puffer, *Takifugu niphobles* (Tetraodontidae). *Environmental Biology of Fishes* 40, 255–261.

Yamahira, K. (1996). The role of intertidal egg deposition on survival of the puffer, *Takifugu niphobles* (Jordan et Snyder), embryos. *Journal of Experimental Marine Biology and Ecology* 198, 291–306.

Yamahira, K. (1997). Hatching success affects the timing of spawning by the intertidally spawning puffer *Takifugu niphobles*. *Marine Ecology Progress Series* 155, 239–248.

5 Catching a Breath: Beach-Spawning Fishes and Air Breathing

By definition, fishes are aquatic vertebrates. The phrase "like a fish out of water" implies a creature that is completely out of place, possibly stressed and confused. The great anthropologist and essayist Loren Eiseley described the terrestrial emergence of a primitive fish as "a monstrous penetration of a forbidden element" (Eiseley, 1946), an excursion filled with terror and aching for breath. Yet many species of beach-spawning fishes emerge from water and are amphibious to some degree. Most of these do so predictably, with deliberate behavior and little outward evidence of fear. Some are so amphibious that they spend up to 90 percent of their time out of water (Clayton & Vaughan, 1986; Sayer & Davenport, 1991; Brown et al., 1992).

During beach spawning, many species of fishes from multiple lineages may briefly emerge from water. Even so, these adults do not necessarily breathe air. Emergence during beach spawning typically occurs at high tides, unlike the more routine amphibious emergence of intertidal fishes during low tides. This chapter addresses amphibious behavior and ability to breathe air for beach-spawning fishes.

5.1 HOW ARE EMERGENCE AND AIR BREATHING BENEFICIAL FOR BEACH-SPAWNING FISHES?

Consider that during a spawning run, large numbers of active fishes congregate in a small area. A high tide temporarily submerges the edge of the coast under shallow water, allowing beach-spawning fishes access to habitats or substrates that may have been previously far inland from the water's edge (Tewksbury & Conover, 1987). Remember that the dissolved oxygen that supports metabolic activity is much lower by volume in water even when in equilibrium with the atmosphere, and that diffusion of oxygen is much slower in water than in air. See Chapter 2 for more details about the physical conditions of the intertidal zone. Thus, although waters of the open ocean rarely become hypoxic (Graham et al., 1978), local conditions indicate a situation in which aquatic oxygen may be insufficient to meet the high demand.

Of course, many beach-spawning fishes do not emerge during spawning but remain in water (DeMartini, 1999; Martin & Swiderski, 2001). For example, the Atlantic silverside *Menidia menidia* (Atherinopsidae) spawns aquatically during high tides over sea grass beds near shore (Middaugh, 1981; Middaugh et al., 1981). The large numbers of fish in the spawning aggregation and their high levels of activity deplete the oxygen in the shallow water, and the area suffers local hypoxia (Figure 5.1).

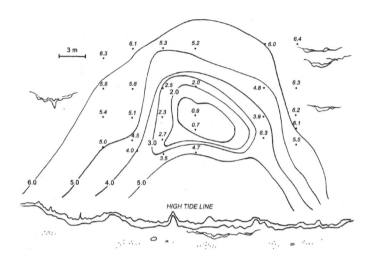

FIGURE 5.1 Oxygen tension in a spawning site for *Menidia menidia*. (After Middaugh, D., *Copeia,* 766–776, 1981. With permission. Artwork by Andrea Lim.)

Adults may become temporarily sluggish and unable to avoid predators during the "spawning stupor" that follows (Middaugh, 1981). One might hypothesize that there would be adaptive benefit for this species to emerge into air for spawning activity if they could tolerate it, rather than becoming incapacitated by the increasingly hypoxic water. Upon return to the water, they could quickly swim away from the spawning area into deeper water and avoid local depletion of oxygen.

5.2 INTERTIDAL FISHES SHOW GRADIENTS OF AMPHIBIOUS BEHAVIOR AND AIR BREATHING

Many marine fishes that reside in the rocky or muddy intertidal zones are amphibious to some degree (Graham, 1976, 1997; Martin & Bridges, 1999; Ishimatsu & Graham, 2011; Martin, 2013). Partially or fully emerging from water, especially during low tides, these fishes breathe air during terrestrial sojourns but not while in the water (Graham, 1976; Bridges, 1988; Martin, 1995; Graham & Lee, 2004). In contrast, most air-breathing fishes that live in freshwater do not emerge onto land but remain aquatic, lifting up the head to the surface to gulp a breath of air while swimming.

Adaptations for air emergence tend to follow a vertical gradient for intertidal fishes at different tidal heights within the rocky intertidal zone (Zander, 1972; Horn & Riegle, 1981; Martin, 1996; Zander et al., 1999; Mandic et al., 2009a). Vertical zonation of marine life is well known in the rocky intertidal zone (Stephenson & Stephenson, 1949; Benson, 2002), in response to different periods of exposure and inundation by tides. The higher the location in the intertidal zone, the longer that site will spend emerged into air during a low tide. Conversely, a site very low in the intertidal zone will spend the greater proportion of time underwater, with emergence only occurring during the most extreme low tides. Comparisons of adaptations for air emergence within one family have provided a model for phylogenetic comparative studies in many intertidal fish species (Zander, 1972; Horn & Riegle, 1981; Low et al., 1990; Martin, 1996; Ip et al., 2002; Ishimatsu & Gonzales, 2011; Richards, 2011).

An intertidal fish, when emerged, may not be immediately apparent to an observer. Most fishes in tide pools are small and cryptic in color and may be difficult to see even when out of water, as they may hide under a rock or in vegetation (Horn & Gibson, 1988; Horn & Martin, 2006). Negatively buoyant and relatively inactive, these amphibious fishes typically lack a gas bladder as adults. They also typically lack organs specifically evolved to breathe air, called air-breathing organs or ABOs. Instead these fishes respire in air with the same structures that are also used for aquatic respiration. See Section 5.3.

It might be suggested that intertidal fishes survive tidal air exposure not because they are specifically adapted to emergence but simply because they are sluggish, inactive fishes with low metabolic demands that can just wait out a tidal cycle. It is true that many intertidal amphibious fishes are relatively inactive when out of water (Ralston & Horn, 1986; Horn & Gibson, 1988). However, laboratory studies indicate that intertidal fishes possess adaptations for emergence that are not present in closely related subtidal fishes (Congleton, 1974, 1980; Graham, 1976; Martin, 1996; Yoshiyama & Cech, 1994; Yoshiyama et al., 1995).

Many species of intertidal amphibious fishes have the capacity for aerobic respiration in air over many hours during tidal exposure, exchanging both oxygen and carbon dioxide at levels similar to aquatic metabolic rates (Teal & Carey, 1967; Graham, 1976; Riegle, 1976; Bridges, 1988; Edwards & Cech, 1990; Martin, 1991, 1993). Most beach-spawning fishes that are resident in the intertidal zone probably have some ability to breathe air when emerged (Martin, 1995; Halpin & Martin, 1999; Luck & Martin, 1999). On the other hand, many beach-spawning species, particularly subtidal residents, briefly emerge from the water during spawning but apparently do not breathe air as adults (Templeman, 1948; Yamahira, 1996; Martin et al., 2004).

For intertidal fishes, three major types of amphibious emergence behavior have been described: passive remainers, tide pool emergers, and skippers (Martin, 1995, 2013). Examples are shown in Figure 5.2 and described below.

Passive remainers reside in the intertidal zone and stay at their usual home shore height during all tides, whether submerged or not. These species are regularly and predictably found either partially emerged in shallow pools or fully emerged from water under boulders or in vegetation on rocky shelves during low tides, simply because they do not leave when the tide ebbs (Martin, 2013). Although relatively quiet and inactive while out of water, they may be out of water for several hours without appearing to be distressed, and they do not seem to seek a rapid return to water. Respirometry shows that they are exchanging both carbon dioxide and oxygen with air during this time. Examples include many species of stichaeids (Horn & Riegle, 1981; Daxboeck & Heming, 1982; Martin, 1995), some blennids (Zander, 1983; Bridges, 1988), gobeisocids (Eger, 1971; Ebeling et al., 1970), and tripterygiids (Innes & Wells, 1985; Hill et al., 1996). This strategy may help these small fishes avoid large predators they would otherwise encounter if they moved with the tides out into the open ocean. Passive remainers also include species with parents that remain during low tides with a nest or clutch of eggs, including some that are not normally resident in the intertidal zone (Crane, 1981; Coleman, 1992, 1999). See Chapters 3 and 4 for more about parental care in beach-spawning fishes.

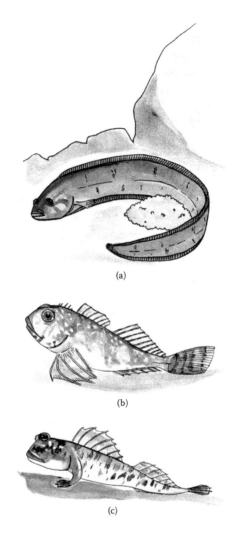

(a)

(b)

(c)

FIGURE 5.2 Vertical height in the intertidal zone is correlated with the type of amphibious behavior and air emergence. Three examples are shown: (a) the intertidal remainer is
Xiphister atropurpureus, a stichaeid fish guarding an egg clutch in the low intertidal zone;
(b) the tide pool emerger is *Oligocottus maculosus*, the tide pool sculpin in the middle intertidal zone; and (c) the skipper is *Periophthalmus modestus*, a mudskipper fully emerged and
active in the supralittoral zone above the water line. (Artwork by Andrea Lim.)

Passive remainer fishes generally are calm and well able to exchange respiratory
gases in air at similar rates to their respiration in water (Riegle, 1976; Daxboeck &
Heming, 1982; Martin, 1993, 2013; Graham, 1976, 1997). In comparison with other
intertidal or subtidal species, those not typically found out of water struggle to survive the few hours of tidal exposure when experimentally stranded (Congleton, 1980;
Davenport & Woolmington, 1981; Martin, 1996; Mandic et al., 2009a). Fishes found
in the lowest edge of the intertidal zone rarely face exposure to air; however, many

of these species tolerate accidental stranding by exchanging respiratory gases in air, even though they may not actively emerge (Martin, 1993; Yoshiyama & Cech, 1994; Luck & Martin, 1999; Zander et al., 1999).

Another type of amphibious behavior by intertidal fishes is seen in active emergers (Martin, 2013). Fishes of these species are not typically found emerged in nature except during hypoxic conditions (Wright & Raymond, 1978). During a low tide, aquatic conditions may change in an isolated pool, particularly during low tides that occur at night or in caves. Hypoxia may occur along with high acidity due to accumulation of dissolved carbon dioxide from respiration (Wright & Raymond, 1978; Truchot & Duhamel-Jouve, 1980). (Review Chapter 2 for details.) When this happens, some tide pool resident fishes choose to actively leave the water, emerging either partially with just the head or moving the whole body fully into air (Wright & Raymond, 1978; Martin, 1991, 1995; Yoshiyama et al., 1995; Mandic et al., 2009a). These active emerger fishes are able to respire in air, taking advantage of the abundant atmospheric oxygen that does not change with the tidal cycle (Congleton, 1980; Davenport & Woolmington, 1981; Martin, 1991, 1993; Watters & Cech, 2003). They can be induced to emerge in the laboratory with hypoxic conditions, perhaps after engaging in aquatic surface respiration (Martin, 1996, 1991; Yoshiyama et al., 1995; Sloman et al., 2008).

Active emergers include many species of Cottidae (Wright & Raymond, 1978; Congleton, 1980; Martin, 1991, 1993, 1996; Yoshiyama et al., 1995; Watters & Cech, 2003; Sloman et al., 2008; Mandic et al., 2009a), Trypterygiidae of New Zealand (Innes & Wells, 1985), Gobiidae (Nilsson et al., 2007; Ishimatsu & Gonzales, 2011) and intertidal Blennidae (Zander, 1972; Bridges, 1988; Zander et al., 1999). Most clingfishes emerge as passive remainers, but the Chilean clingfish *Sicyases sanguineus* (Gobiesocidae) has been reported to move ahead of the tide line to resist return to water (Ebeling et al., 1970; Gordon et al., 1970). All of these species show active avoidance of hypoxia as adults, although many also have a great tolerance for low aquatic oxygen. Many of these species are known to nest so that their eggs are emerged by low tides into air (Coleman, 1999; DeMartini, 1999).

The third type of amphibious emergence behavior for intertidal fishes is the highly terrestrial skippers. Rockskippers and mudskippers are highly amphibious fishes that actively emerge from water for most of their daily activities (Graham & Rosenblatt, 1970; Zander, 1972, 1983; Graham et al., 1985; Brown et al., 1992; Clayton, 1993). This type of emergence seems to be restricted to only a few fish families—Gobiidae, Blenniidae, and Labrisomidae—but each has multiple amphibious species (Graham, 1997; Martin, 2013). Also within each of these families are gradients of adaptation, with some amphibious species that are passive remainers and others that are tide pool emergers, as well as some species that are completely aquatic.

Rockskipper blennids and labrisomids occupy the supralittoral zone of rocky shores in the tropics and are well adapted for air emergence (Graham et al., 1985; Martin & Lighton, 1989; Brown et al., 1992; Zander et al., 1999; Bhikajee & Green, 2002; Neider, 2005). Mudskippers are tropical and temperate Oxudercine gobies found mostly in the Indo-Pacific (Clayton, 1993). They are engagingly active out of water on mud flats during lower tides, whenever terrestrial habitat is available (Tamura et al., 1976; Low et al., 1990; Randall et al., 2004). Although not every species has been studied extensively, many are known to spawn in burrows or crevices at the water's edge.

Both mudskippers and rockskippers are very well adapted for air emergence and amphibious life. Feeding, displays for territoriality and reproduction, spawning, and parental care occur out of water (Clayton & Vaughan, 1986; Clayton, 1993; Graham, 1997; Shimizu et al., 2006; Ishimatsu et al., 2007; Pace & Gibb, 2009; Hsieh, 2010). Skipper species emerge from normoxic water as well as hypoxic water as part of daily life, not necessarily as an escape from unpleasant aquatic conditions. They never stray far from a source of liquid water, however, and have characteristic behaviors to rehydrate respiratory membranes by taking a frequent quick sip or dip (Ip et al., 1991; Brown et al., 1992). They never make the kinds of extensive overland excursions seen in some freshwater amphibious fishes (Hensley & Courtenay, 1980; Graham, 1997; Graham & Lee, 2004).

Mudskippers dig mud burrows that hold water even during an ebb tide. When the tide goes out, these fishes must either emerge into air or submerge into aquatic hypoxia. In the mud, anoxia occurs within a few centimeters from the benthic surface. The adults take refuge within mud burrows to avoid desiccation, and some species carry mouthfuls of air to fill a space within the chamber. When nesting, the air surrounds their egg clutches during incubation (Ishimatsu et al., 2007, 2009). For more on this intriguing method for avoiding aquatic hypoxia and its impact on the embryos, see Chapters 7 and 8.

Skippers are known from tropical and temperate habitats, while passive remainers and tide pool emergers are known mainly from temperate waters (Martin, 2013). The increased tendency for emergence in tropical intertidal fish species may arise from the effects of temperature on ectotherms, as higher temperatures both increase metabolic demand and decrease the solubility of oxygen in seawater (Dejours, 1994; Martin & Bridges, 1999; see Chapter 2).

5.3 RESPIRATORY STRUCTURES FOR AMPHIBIOUS FISHES ARE SIMILAR FOR WATER AND AIR

Marine amphibious intertidal fishes use the same respiratory structures to exchange oxygen and carbon dioxide in both air and water. These fishes typically lack a swim bladder and do not have a unique respiratory organ for breathing air (Graham, 1976, 1997). Instead, these fishes use gills, skin, and in some cases highly vascularized areas of the bucco-opercular mucosae for respiratory gas exchange of oxygen and carbon dioxide in both air and water (Low et al., 1990; Graham, 1997; Gonzales et al., 2011). In contrast, freshwater air-breathing fishes typically have some sort of specialized air-breathing organ (ABO), such as a lung, some special vascularization of an area of the digestive tract, or an elaborate infolding of the opercular chamber. For extensively illustrated reviews of ABOs, see Randall et al. (1981), Graham (1997), and Ishimatsu (2012).

Using the gills and skin for aerial respiration seems counterintuitive, but it has at least one advantage over use of an enclosed ABO or lung (Randall et al., 1981; Feder & Burggren, 1985; Dejours, 1994). Gills provide a one-way flow-through respiratory system, while the lungs require tidal exchange of air that may be unable to completely empty in order to fully exchange all the air within it (Piiper & Scheid, 1975).

Gills are filamentous respiratory structures derived from embryonic ectoderm. They are in direct contact with the respiratory medium (Piiper & Scheid, 1975), and water is pumped in only one direction over their surfaces, from the mouth across the gills to exit through the opercular opening (Figure 5.3a). On the other hand, lungs are enclosed respiratory structures derived from embryonic endoderm, with tidal flow of air in and then back out through the mouth in fishes, or the nose in tetrapods (Figure 5.3b). This two-way flow causes some mixing of fresh air with residual air that remains in the lung or ABO, even after exhalation, so that the respiratory medium within the lung usually has lower oxygen and higher carbon dioxide tensions than the surrounding atmosphere. Since atmospheric air has so much more oxygen per unit volume than water does, this decrease is a manageable constraint, given the increased protection from desiccation obtained by enclosing the respiratory organ.

Even so, a residual level of pulmonary carbon dioxide may build up and result in respiratory acidosis (Graham, 1976, 1997). If an air-breathing fish uses the lung primarily for oxygen exchange while swimming, it can get around this constraint by expelling excess carbon dioxide into the water through a different respiratory organ, via the gills or skin (Graham, 1976; Feder & Burggren, 1985). However, when out of water and only able to use the lung or ABO, burrowing lungfish *Protopterus aethiopicus* cannot release carbon dioxide at the same rate as it is produced, and they develop respiratory acidosis and blood hypercarbia (Johansen & Lenfant, 1968; Delaney et al., 1974).

For any animal, respiratory organs must have sufficiently large surface area to allow diffusion of oxygen in and carbon dioxide out, and a moist membrane in

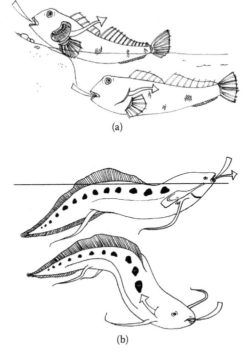

(a)

(b)

FIGURE 5.3 Comparison of (a) one-way flow of water over gills with (b) tidal flow of air in lungs. (Artwork by Andrea Lim.)

contact with the respiratory medium (Piiper & Scheid, 1975; Feder & Burggren, 1985). In water, desiccation of respiratory surfaces is not an issue, but in air, a conflict exists between respiratory gas exchange and water loss by evaporation (Dall & Milward, 1969; Horn & Riegle, 1981; Luck & Martin, 1999). Decreasing the moist respiratory surface area of the gills reduces the potential for desiccation and also minimizes the difference in gill surface area that occurs during air exposure for most fully aquatic fishes, when the long, delicate gill filaments collapse.

In water, highly elaborate gill filaments provide a large surface area and moisture is omnipresent. In air, however, gills tend to collapse, reducing surface area for diffusion and gas exchange. Many amphibious fishes show loss of gill surface area, indicating that cutaneous respiration may become more important for amphibious fishes (Jew et al., 2013). An adaptive reduction in gill surface area, as compared with fully aquatic fishes of similar size, is usually made by decreasing the length and increasing the width of the filaments. This is most pronounced in the highly terrestrial mudskippers (Low et al., 1990; Graham, 1997; Gonzales et al., 2011; Ishimatsu, 2012). However, it is the opposite anatomical response that one sees for fishes faced with chronic hypoxia, when gill filaments become longer and more extensive (Sollid & Nilsson, 2006; Tzaneva et al., 2011).

The reduction in gill surface area of amphibious intertidal fishes indicates an anatomical trade-off between an increased ability to breathe air and a decreased capacity for exchanging respiratory gases in water (Martin & Bridges, 1999). This conflict sets the stage for the necessity of increased physiological tolerance to hypoxia that is seen in many tide pool fishes (Richards et al., 2008; Sloman et al., 2008; Richards, 2011; Craig et al., 2014). Decreased surface area of the gills helps prevent collapse when the fish is in air, but reduces the area for diffusion of respiratory gases (Graham, 1997). Since many intertidal fishes are relatively inactive, this loss of capacity may not be a great constraint. Even the most amphibious skipper fishes may show very low activity when in water (Ishimatsu & Gonzales, 2011).

In addition to protection against collapse, decreased gill surface area may reduce the surface area for desiccation in species that emerge into air (Low et al., 1990; Graham, 1997; Gonzales et al., 2011). Otherwise, the one-way flow of air over the gills through the mouth and opercular chambers can lead to rapid desiccation of respiratory membranes. Some mudskippers can close the opercular chamber during terrestrial excursions and hold a small amount of water inside (Graham, 1997; Aguilar et al., 2000). Rather than using this water for respiratory oxygen, it appears that this tiny puddle is used to maintain gill hydration by sloshing around (Graham, 1997). The sealed opercular chambers can also be used by mudskippers *Periophthalmus modestus* and *Periophthalmodon schlosseri* to deliver mouthfuls of air to underwater mud burrows to protect incubating embryos from aquatic hypoxia (Ishimatsu et al., 2007, 2009).

A lack of scales is common among intertidal fishes, but cutaneous blood vessels can be found in the epidermis above dermal scales in many fishes (Gonzales et al., 2011). Modifications of skin morphology such as lack of scales and increased vascularization may aid cutaneous respiratory gas exchange (Feder & Burggren, 1985; Zhang et al., 2000; Gonzales et al., 2011). Blood vessels around the head of some skippers redden during air exposure (Todd & Ebeling, 1966; Zander, 1972; Zander et al., 1999), indicating increased blood flow to the skin.

When intertidal fishes are out of water, they must keep the gills and skin surfaces moist. This they can accomplish even with a very shallow water source. Their behaviors include rolling on the side in a shallow pool and quick dips and jumps in and out of water (Ip et al., 1991; Brown et al., 1992). The rockskipper *Alticus kirkii* can pump water over the gills if just the mouth is in contact with water, even if the rest of the head and body are emerged. Water moves up through the mouth from shallow depressions, to travel over the gills and spill out the opercular openings (Brown et al., 1992).

Amphibious marine fishes never venture far from their home pool or burrow. Microhabitat choices made by amphibious intertidal fishes during tidal emergence reduce exposure to desiccation and hypoxia. Besides returning to pools regularly, fishes shelter under boulders or within crevices, often near a small amount of seawater that remains after the tide recedes. Whether or not the water becomes hypoxic, if a fish can maintain adequate moisture on the gills to avoid desiccation, it can access the unlimited reservoir of oxygen in the atmosphere (Graham, 1997; Martin & Bridges, 1999). These microhabitats, damp but air-exposed at low tides, are ideal for beach-spawning fishes and their nests (Crane, 1981; Marliave & DeMartini, 1977; Martin et al., 2004; DeMartini, 1999).

5.4 RESPIRATORY STRUCTURES UNDERGO MODIFICATION AS CONDITIONS CHANGE

Modifications of the gills to improve aerial respiration in amphibious fishes appear to be correlated with the level of terrestrial activity and the duration of terrestrial exposure (Low et al., 1990; LeBlanc et al., 2010; Gonzales et al., 2011). Some amphibious fishes show structural cartilage within or between the primary gill filaments to increase support when out of water (Low et al., 1990; Graham, 1997; Gonzales et al., 2011).

The most highly modified respiratory structures of amphibious fishes are seen in those that are either the most active when emerged, or those that may be emerged for long periods of time (Low et al., 1990; Graham, 1997; Zhang et al., 2000; Gonzales et al., 2011). The gobiid *Pseudapocryptes lanceolatus* lives on mud flats, breathes air, and shows reduced gill filaments (Yadav & Singh, 1989). Temperature and ambient oxygen influence gill remodeling in water as in air (Sollid & Nilsson, 2006), and there may be plastic responses at different times of exposure as for *Kryptolebias marmoratus* (LeBlanc et al., 2010).

5.5 METABOLIC RATE IN AIR DURING EMERGENCE IS SIMILAR TO METABOLIC RATE IN WATER FOR MANY AMPHIBIOUS FISHES

In air, the metabolic rates of most amphibious intertidal fishes are equivalent to their aquatic metabolic rates (Graham, 1976, 1997; Bridges, 1988; Martin & Bridges, 1999). The possibility that some amphibious fishes have reduced metabolic rates out of water has been suggested for several intertidal species, including the mudskipper *Periophthalmus argentilineatus* (Garey, 1962), the California grunion *Leuresthes tenuis* (Garey, 1962), a rockskipper *Alticus kirki* (Brown et al., 1992), and the clingfish *Sicyases sanguineus* (Gordon et al., 1970). This may be a kind of reverse diving response, decreasing metabolism to conserve energy, as indicated by decreases in heart rate and opercular ventilation rate.

Respiratory gases follow partial pressure gradients for diffusion, and oxygen consumption by metabolism maintains a high gradient for influx of oxygen into animals' blood whether they are in water or in air. As a respiratory medium, air is much less viscous than water. Oxygen makes up a much higher portion of the volume of air than of the same volume of water, even when the partial pressures for both water and air in equilibrium are equal (Dejours, 1994). See Chapter 2 for more details.

For all amphibious fishes, the rate of gill ventilation by opercular pumping decreases dramatically in air because of the differences in viscosity, diffusion rate, and oxygen content between air and water (Martin & Bridges, 1999). Differences in oxygen volume within water as compared to air even at equilibrium at the same tension allow many fishes to have a much slower tempo of opercular movement in air than in water (Bridges, 1988), while still maintaining the same rate of oxygen consumption. Intertidal fishes respiring aquatically cope with the low solubility and slow diffusion rate of oxygen by pumping the operculae faster than when breathing in air (Martin & Bridges, 1999). Some fishes may cease opercular movements or seal up the opercular border when emerged into air (Graham, 1997; Martin & Bridges, 1999).

The only accurate way to compare metabolism in water with metabolism in air is to measure the actual consumption of oxygen and release of carbon dioxide by respirometry (Graham, 1997). Unfortunately, methods used to measure dissolved oxygen in water are different from those used to measure the oxygen contents of air. Separate studies of the same species under different conditions are not necessarily equivalent, so it is difficult to generalize. The mudskipper *Periophthalmus argentilinatus* ventilates intermittently, moving its operculae less than half the time while in normoxic water, and the shanny *Blennius pholis* changes both depth and frequency of opercular movements during air exposure (Pelster et al., 1988). Both of these factors could affect the time needed to measure metabolic rates accurately. Another confounding variable is the potential for anaerobic metabolism by intertidal fishes in air (Martin, 1996).

Carbon dioxide produced metabolically is much more soluble in water than oxygen and can easily diffuse out across the gills following its partial pressure gradient when fishes are in air-equilibrated water (Dejours, 1994). However, during nocturnal low tides, aquatic hypercarbia and lower pH can occur in tide pools when many animals and plants are respiring but no photosynthesis is replacing oxygen or removing carbon dioxide (Truchot & Duhamel-Jouve, 1980). If an amphibious fish emerges from the water to breathe air, the same respiratory surfaces that exchange oxygen aerially can also exchange carbon dioxide. Because the gills and skin perform this function, and they are in direct contact with the respiratory medium in both water and air (Piiper & Scheid, 1975), the exchange of carbon dioxide follows its partial pressure gradient to be readily released into air (Martin, 1993, 1995; Martin & Bridges, 1999), in contrast to lungfishes on an enclosed ABO in air during aestivation, particularly if the skin is prevented from gas exchange by enclosure in a cocoon (Delaney et al., 1974). Thus amphibious fishes reduce the likelihood of respiratory acidosis during terrestrial emergence.

5.6 TERRESTRIAL ACTIVITY INCREASES THE RATE OF AERIAL GAS EXCHANGE FOR AMPHIBIOUS FISHES

Most intertidal fishes do not engage in sustained levels of high activity, whether in or out of water (Ralston & Horn, 1986; Horn & Gibson, 1988; Horn & Martin, 2006). Even the black prickleback *Xiphister atropurpureus*, a passive remainer stichaeid, can double its rate of oxygen consumption in air over its resting aquatic oxygen consumption (Daxboeck & Heming, 1982). The high rate of oxygen consumption of *X. atropurpureus* in air that was measured by Daxboeck and Heming (1982) was considered a "resting" rate. However, later measurements with different equipment found an aerial resting metabolic rate for *X. atropurpureus* that was lower, equal to the aquatic metabolic rate (Martin, 1995). The high oxygen consumption rate measured in air for this species suggests that aerobic scope for activity exists in amphibious fishes, even in species that are relatively inactive out of water most of the time. This scope could assist escape from terrestrial predators.

Mudskippers and rockskippers apparently power their usual high levels of terrestrial activity primarily with aerobic metabolism, without incurring an oxygen debt (Graham et al., 1985; Hillman & Withers, 1987; Martin & Lighton, 1989; Kok et al., 1998; Jew et al., 2013). During air exposure, amphibious fishes are less likely to need assistance from anaerobic metabolism than closely related, fully aquatic species from the subtidal zone (Martin, 1996). Upon return to water after a period of time emerged into air, even passive remainer fishes quickly resume the same aquatic metabolic rate as before emergence, indicating a lack of oxygen debt (Edwards & Cech, 1990; Martin & Bridges, 1999; Luck & Martin, 1999).

Oxygen consumption during periods of high activity has been examined for only a few skipper types of intertidal fishes, the mudskipper *Periophthalmus argentilinatus* (Hillman & Withers, 1987), the rockskipper *Alticus kirki* (Martin & Lighton, 1989), the barred mudskipper *Periophthalmus modestus*, and the giant mudskipper *Boleophthalmus boddaerti* (Kok et al., 1998; Jew et al., 2013). These highly terrestrial fishes are able to greatly increase their oxygen consumption rates during activity in air, well above resting levels.

Aerobic metabolism appears to be sufficient to power most terrestrial activity bouts for skippers, although anaerobic metabolism may be called upon if needed (Hillman & Withers, 1987; Martin & Lighton, 1989; Jew et al., 2013). Carbon dioxide release in *A. kirki* increases to keep pace with the increase in oxygen consumption during activity (Martin & Lighton, 1989). In fishes, rapid bouts of exercise that are fueled anaerobically may incur recovery times of many hours because of accumulation of lactate (Teal & Carey, 1967; Riegle, 1976).

An oxygen debt following anaerobic activity is shown by an increased oxygen consumption rate over the resting rate, to convert lactate back into glucose, a process that requires energy input. This is followed by a return to resting metabolic rate. Recovery after terrestrial activity is faster in air than in water for the skipper *Periophthalmus modestus*, while recovery is faster in water than in air for the less amphibious *Boleophthalmus boddaerti* (Kok et al., 1998), revealing another gradient of adaptations for aerial emergence and activity on land between species. Recovery from terrestrial activity can be accelerated in air for *P. modestus* by artificially increasing the oxygen content of the atmosphere, indicating that the respiratory system has functional reserves (Jew et al., 2013).

5.7 MANY AMPHIBIOUS FISHES SHOW HYPOXIA TOLERANCE

In most fishes, ventilation rate and gill filament remodeling respond to ambient oxygen tensions. These oxygen tensions are far more variable in water than in air (Truchot & Duhamel-Jouve, 1980; Randall et al., 1981; Dejours, 1994). During aquatic hypoxia, some fishes increase ventilation rate and swim near the surface (Congleton, 1980; Yoshiyama et al., 1995; Martin, 1996; Sloman et al., 2008). Some species gulp a bubble of air in the mouth for buoyancy, which may also be a source of respiratory oxygen (Gee & Gee, 1994). The simple behavior of leaving the water, if possible, may be more straightforward and more advantageous for some species than making physiological accommodation to poor aquatic conditions. Mobile adults may be able to avoid or recover from hypoxia exposure by escaping from a hypoxic pool, either emerging into air or leaping into a different pool of well-aerated water nearby.

Undoubtedly, many intertidal fishes employ multiple strategies to maintain metabolism in a changeable environment. Tolerance of hypoxic aquatic conditions may be extremely important for some intertidal fishes, especially mudskippers and other gobies that nest in mud burrows (Ishimatsu et al., 2007, 2009). Tolerance to hypoxia is seen in males of the beach-spawning midshipman *Porichthys notatus* (Craig et al., 2014), found in shallow pools under boulders during low tides while they guard nests. These adults decrease metabolism and tolerate the build-up of lactate under hypoxic conditions, and also have an increased hematocrit to store oxygen and assist its transport (Craig et al., 2014). This species is also able to breathe air when emerged by a low tide (Martin, 1993).

The sculpins (Cottidae) show a gradient of adaptation for air emergence and gas exchange in air, and also show a variety of biochemical responses to aquatic hypoxia (Richards, 2011). Cottids compared across an intertidal habitat showed that an increased ability to take up oxygen from water provided greater tolerance to aquatic hypoxia. The tide pool sculpin *Oligocottus maculosus* changes gill surface area, increases affinity of hemoglobin for oxygen, and decreases metabolic rate during hypoxia (Sloman et al., 2008; Mandic et al., 2009b). Both *O. maculosus* and the fluffy sculpin *O. snyderi* produce heat shock proteins (HSPs) for thermal challenges during air emergence (Nakano & Iwama, 2002). These HSPs may also be produced in response to hypoxia rather than heat stress (Iwama et al., 2004) and may aid in tolerance to multiple stressors including changing temperatures, osmotic stress, and hypoxia (Todgham et al., 2005; McBryan et al., 2013).

5.8 ESTUARINE FISHES THAT SPAWN ON BEACHES MAY BREATHE AIR WHETHER OR NOT THEY EMERGE

Coastal estuaries are tidally influenced but typically retain areas of open water during low tides, so resident estuarine fishes are less likely to be emerged into air on a regular basis than are resident fishes from the rocky or muddy intertidal habitats. However, because of the low wave action, estuarine waters experience aquatic hypoxia with a greater frequency than tide pools. Adaptations for air emergence and tolerance for aquatic hypoxia certainly would be adaptive for estuarine fishes and their nests.

Many teleost fishes living in estuarine habitats are able to breathe air in response to hypoxia, but may do so in a manner very different from the amphibious intertidal fishes discussed previously. For example, the family Gobiidae has many estuarine species, including mudskippers. The longjaw mudsucker *Gillichthys mirabilis* has occasionally been observed out of water (Todd, 1968). It tolerates hypoxic water well (Gracey et al., 2011). In hypoxic water, it gulps air at the surface, probably for gas exchange at the gills. Its reddened head and lips reveal the extensive vascularization of its cranial cutaneous surfaces that may also contribute to respiratory gas exchange (Todd & Ebeling, 1966).

Another gobiid, the tidewater goby *Eucyclogobius newberryi* has been observed in anoxic water gulping air using aquatic surface respiration (C. Swift, personal communication). This protected species nests in sandy mud burrows and provides male parental care (Swenson, 1997). How the adults and embryos survive incubation in burrows is not known, but parental care seems to be necessary for the eggs to survive. See Chapter 3 for more on reproduction of this species.

Killifish (Fundulidae) are common cyprinodonts in temperate and tropical estuaries. Many species have eggs that are emerged into air during some or all of their incubation period (Taylor, 1984; Martin, 1999). Hypoxia tolerance causes changes in carbohydrate metabolism for the Gulf killifish *Fundulus grandis* (Martinez et al., 2006). The killifish *Fundulus heteroclitus* rarely emerges from water in its estuarine habitat but can breathe air (Halpin & Martin, 1999). This species also is well adapted to aquatic hypoxia and adjusts carbohydrate metabolism to use less oxygen by regulating pyruvate dehydrogenase (Richards et al., 2008). Many killifish spawn on beaches (see Chapter 3), but few have been studied for air-breathing or amphibious adapations.

The mangrove killifish *Kryptolebias marmoratus* (Rivulidae) survives extended aerial emergence of weeks or months, taking refuge in rotting tree trunks and vegetation (LeBlanc et al., 2010). In air, the gills of *K. marmoratus* undergo extensive remodeling that reduces gill surface area by filling in the interlamellar spaces (Wright, 2012). The skin takes on more respiratory function in air through angiogenesis. This species may emerge into air in response to aquatic hydrogen sulfide (Abel et al., 1987) as well as hypoxia. When aerial exposure is prolonged beyond a few hours of a tidal cycle, morphological effects increase (LeBlanc et al., 2010), including a profound reorganization of the gill and skin structure during periods of terrestrial life that may go on for months (Wright, 2012).

5.9 FISHES THAT MIGRATE INTO THE INTERTIDAL ZONE AND EMERGE DURING BEACH SPAWNING DO NOT BREATHE AIR

Subtidal species that spawn on beaches include species from the Tetraodontidae, the kusafugu puffer *Takifugu niphobles* (Yamahira, 1996), four Osmeridae, the capelin *Mallotus villosus* (Templeman, 1948), two species of surf smelt (Saruwatari et al., 1997), *Hypomesus pretiosus* (Hart, 1973) and *H. japonicus* (Hirose & Kawaguchi, 1998), and night smelt *Spirinchus starksi* (Sweetnam et al., 2001). Four atherinopsids, the Atlantic silverside *Menidia menidia* discussed at the start of this chapter (Middaugh et al., 1981), the California grunion *Leuresthes tenuis* (Walker, 1952), its congener the Gulf grunion *L. sardina* (Thomson & Muench, 1976), and the false grunion *Colpichthys regis* (Russell et al., 1987) live subtidally but spawn in the intertidal zone. The Pacific herring *Clupea pallasii pallasii* (Clupeidae) spawns over intertidal sea grass beds in great numbers (Flostrand et al., 2009). These fishes in most cases are emerged only briefly and perhaps accidentally during spawning, and most if not all apparently do not breathe air to any great extent while emerged. None of these remain guarding the embryos after oviposition. However, little research has been done on aerial respiration in these species.

Some beach-spawning fishes may reduce metabolic rate or breath-hold while in air (Garey, 1962; Martin et al., 2004). The atherinopsid *Leuresthes tenuis,* the California grunion, comes completely out of the water to spawn and may remain out of water for several minutes, moving about on shore (Walker, 1952; Martin et al., 2004). Spawning involves coordinated activity including surfing onto shore on a large wave, then jumping or wriggling some distance across the sand. Females burrow tail first into the soft sand to release eggs under the surface while the males curl about them, providing milt. The heart rate of *L. tenuis* declines when these fish are artificially emerged into air (Garey, 1962). However, the normal spawning activity for both male and female *L. tenuis* is accomplished aerobically and does not require inputs of anaerobic metabolism (Martin et al., 2004). In spite of its vigorous terrestrial spawning activity, *L. tenuis* does not seem to breathe air and cannot survive much more than a mere 20 minutes out of water, even when calm (Martin et al., 2004). Compare this with some skipper or passive remainer fishes that easily survive emergence of many hours (Martin, 1991, 1995), days, or even weeks (LeBlanc et al., 2010; Wright, 2012).

If forced into terrestrial activity by experimental manipulation, *L. tenuis* is eventually unable to remain upright. Lactic acid accumulates during this exhausting terrestrial activity (Scholander et al., 1962; Martin et al., 2004), indicating anaerobic metabolism was triggered. The contrast between anaerobic metabolism during forced exercise, versus aerobic support of spawning activity during volitional emergence, suggests that this species has a sort of reverse diving response during spawning that reduces metabolic effort in air, rather than adaptations for exchanging oxygen and carbon dioxide in air (Garey, 1962; Scholander et al., 1962; Martin et al., 2004). Most individuals of both species of *Leuresthes* survive their unusual type of spawning emergence, and males and females can spawn repeatedly over the season through several years of life (Walker, 1952; Thomson & Muench, 1976).

In contrast, *Mallotus villosus* behavior during spawning involves crowding close to the water line but not deliberate emergence (Frank & Leggett, 1981). The spawning run is called *capelin rolling* for the tightly wound clusters of fish moving about in shallow water. Following a run, the shoreline may be covered with spent fish that have died, whether by stranding or other cause (Nakashima & Wheeler, 2002). It has been suggested that the males are less likely to survive a spawning run than the females, indicating a gender-specific life-history strategy (Huse, 1998). Unlike most of the other species discussed in this text, *M. villosus* also spawn in deeper water. Some of, their deeper spawning sites appear to be locations of former coastlines that have been inundated by geologic activity (Nakashima & Wheeler, 2002). See Chapter 4 for more details about spawning behavior.

5.10 THE GAS BLADDER HAS NONRESPIRATORY FUNCTIONS IN BEACH-SPAWNING FISHES

The gas bladder or swim bladder of fishes is a derivative of ancient lungs (Graham, 1997). Perhaps it is a bit surprising that there is no known example of a beach-spawning fish species that uses a gas bladder as a lung. In fact, the vast majority of amphibious intertidal fishes do not even have a gas bladder as adults (Graham, 1976; Horn et al., 1999; Horn & Martin, 2006). For those species that have one, the gas bladder or swim bladder of fishes may take either of two different forms during adaptation for different functions.

In the first form, the swim bladder is connected to the esophagus by a tube, open through this structure to the atmosphere. Fishes with an open gas bladder can gulp air at the surface and use the gas bladder for respiratory gas exchange. This structure is sometimes called a lung, and it is homologous to the tetrapod lung (Graham, 1997; Zheng et al., 2011), although in fishes it is usually above or dorsal to the gastrointestinal tract, whereas in amphibians and land vertebrates the lung is below, or ventral to the gastrointestinal tract. Sometimes this structure is called an air-breathing organ, or ABO, and that name can also be applied to many different, nonhomologous respiratory structures found in air-breathing fishes (Graham, 1997). Respiratory structures such as an open or physostomous gas bladder are characterized by thin, well-vascularized membranes, and lungs must be elastic enough to inflate and deflate quickly and smoothly.

The second form of the swim bladder, very common among teleosts, is the closed form, with no duct between the gas bladder and the esophagus. Gases are moved into and out of this hollow physoclistous organ by means of exchange with the blood across capillary beds within the animal. Local chemical environments and multiple hemoglobin types allowing the movement of oxygen from the blood into the gas bladder lumen, a region that may already have high partial pressures of oxygen. Gases can be withdrawn or added, but the process is slow and very gradual, unlike the emptying and filling of a lung. In addition, to maintain the inflation of the gas in the bladder, the membrane within it is thick and tough, with little vascularization except at the two regions where gases are either absorbed or expelled into the space within the bladder. Thus, although there may be high oxygen tensions in gas bladders of some fishes, the oxygen is not readily accessible for respiration. Instead, the gas bladder provides a way to control buoyancy in the water column. This means it has no apparent respiratory function while the fishes are emerged from water.

Physoclistous gas bladders are found in several species of beach-spawning fishes from the subtidal zone, including *Leuresthes tenuis* and *Mallotus villosus*, that probably use it for buoyancy. Some say that *L. tenuis* make a faint sound while spawning (Walker, 1949; Spratt, 1986). This probably would be generated by vibrations in the closed gas bladder, but neither the sound nor its source have been confirmed experimentally.

A nonrespiratory function of a closed gas bladder comes from the beach-spawning midshipman *Porichthys notatus*. Males set up nesting sites under boulders in the intertidal zone and then create humming vibrations and deep tones to attract females to breed (Brantley & Bass, 1994). The noise generated is so impressive that it affects mate choice (Vasconcelas et al., 2011). See Chapter 10 for more on the local human response.

The herring *Clupea pallasii pallasii* has an unusual physostomous swim bladder with two openings. It is used for buoyancy rather than respiration (Blaxter & Batty, 1984) and may also provide a novel means for sound production underwater by controlled release of bubbles (Wilson et al., 2004).

Only one marine teleost breathes air with the assistance of a nonrespiratory swim bladder, the Pacific fat sleeper *Dormitator latifrons* (Todd, 1973). This species is not amphibious but lives near shore estuaries and moves between seawater and freshwater. Its closed swim bladder is a flotation device, allowing it to rise to the surface of hypoxic water for respiratory gas exchange to take place cutaneously through extensive vasculature on the top of the head, held up by the buoyant swim bladder (Todd, 1973; Graham, 1997).

5.11   AIR EMERGENCE HAS PHYSIOLOGICAL CONSEQUENCES FOR FISHES

Desiccation during air emergence can lead to hyperosmolarity of body fluids for fishes, causing impaired cardiovascular and renal functions (Dall & Milward, 1969). Removal of nitrogenous wastes is more difficult in air. Mudskippers have a high tolerance for ammonia during emergence (Evans et al., 1999; Ip et al., 2002; Randall et al., 2004).

The gills and skin of amphibious fishes contain ion channels for permeability and transport systems (Evans et al., 1999; LeBlanc et al., 2010). Loss of gill surface area is adaptive for amphibious fishes that breathe air, but it reduces their capacity for ion transport and may cause increases in ion transport across the skin. Changes in ion transport while emerged may be partly responsible for changes in metabolic rates between water and air, for those amphibious fishes in which this is seen (Uchiyama et al., 2012).

Moisture of skin mucous serves important protective functions in addition to enabling respiratory gas exchange. Prickles and skin ornamentation on cottids may slow water loss. Water lost through evaporation in air may take a long time to replace following return to seawater (Horn & Riegle, 1981; Luck & Martin, 1999).

5.12   SUMMARY AND CONSEQUENCES: ADULTS HAVE GREATER PHYSIOLOGICAL TOLERANCE TO HYPOXIA THAN EMBRYOS; EMBRYOS ARE BETTER SUITED FOR AIR EXPOSURE

Adult fishes that spawn on beaches are rarely emerged, but the eggs and embryos are frequently emerged. It appears that adults may have a combination of strategies for dealing with the challenges of air emergence and aquatic challenges, including decreasing metabolism, tolerance of aquatic hypoxia, and ability to exchange both oxygen and carbon dioxide in air. However these species seem to be more likely to breathe air during the embryonic stages than as adults (Martin, 1999; Martin et al., 2011; Martin & Carter, 2013). See Table 5.1 for a summary of the types of emergence seen in beach-spawning fishes.

TABLE 5.1

Patterns of Terrestrial Emergence for Beach-Spawning Fishes: For Beach-Spawning Fishes, Spawning Usually Takes Place Aquatically during High Tides

Adult Habitat	Spawning Habitat	Adults Emerge at High Tide	Egg Air Emerge	Adults Emerge at Low Tide	Examples	Adult Breathes Air?
Rocky intertidal	Rocky intertidal	No	Frequent, tidal	Tidal	*Xiphister atropurpureus*	Y
					Gobiesox meandricus	Y
					Oligocottus maculosus	Y
Mud flats	Mud flats	No	Constant, air chamber in burrow	Tidal	*Periophthalmus modestus*	Y
Rocky intertidal	Rocky intertidal	Yes, frequently	Daily, tidal	Yes	*Adamia tetradactyle*	Y
Subtidal	Rocky intertidal	No	Frequent, tidal	No	*Gasterosteus aculeatus* ("white")	U
					Clinocottus acuticeps	U
Subtidal	Rocky intertidal	No	Frequent, tidal	Tidal	*Porichthys notatus*	Y
Subtidal	Gravel beach or coarse sand	Accidental	Frequent, tidal	No	*Ammodytes hexapterus*	N
					Takifugu niphobles	N
					Mallotus villosus	N
					Hypomeses pretiosus	N
Subtidal	Sandy beach	Yes, to spawn	Constant, above mean high tide	No	*Leuresthes tenuis*	N
					Leuresthes sardina	N
Estuarine	Vegetation, shells	Accidental	Frequent, tidal	No	*Fundulus heteroclitus*	Y
					Menidia menidia	N

Note: For eggs and adults of these species, emergence takes place only during low tides. The two *Leuresthes* species are exceptional in that spawning adults emerge from water during high tides, and the eggs are emerged (buried in sand) for the entire incubation period. Air-breathing ability is either Y = Yes, N = No, or U = Unknown.

Intertidal fishes stay close to water and do not seem to be able to resist water loss and desiccation (Horn & Riegle, 1981; Luck & Martin, 1999). They generally do not venture shoreward from their intertidal habitat whether feeding or interacting with conspecifics, or trying to avoid predators, although some mudskippers may move up onto mangrove roots or other exposed plants during low tides (Graham, 1976, 1997). No known species of amphibious intertidal fishes can move great distances across terrestrial habitats except during a flood tide (Graham et al., 1985; Hsieh, 2010; Ishimatsu & Gonzales, 2011). Given the current state of their adaptations, these fish seem to be unable to colonize land beyond the intertidal zone (Horn et al., 1999; Graham & Lee, 2004; Horn & Martin, 2006).

Although guarding adults may be able to tolerate temporary aquatic hypoxia in a burrow or nest, rapidly developing embryos within eggs must be positioned to avoid aquatic hypoxia (Taylor et al., 1977; Middaugh et al., 1983; Tewksbury & Conover, 1987; Strathmann & Hess, 1999). Oviposition in the intertidal zone allows eggs to be emerged into air by a low tide, rather than being submerged in stagnant, hypoxic and hypercarbic water in an isolated pool or estuary. Conditions may be improved by attaching the eggs to the ceiling of a rock chamber formed by a boulder over a pool (Crane, 1981; Coleman 1999), or by deliberate introduction of mouthfuls of air into a chamber within a mud burrow by guarding parents (Ishimatsu et al., 2009). Placing the clutch in or on an appropriate substrate also helps avoid desiccation (Taylor et al., 1977; Middaugh et al., 1981; Martin, 1999). See Chapter 7 for more about parental care and other features of terrestrial incubation in beach-spawning fishes.

Predation and other threats to spawning adults and incubating embryos during their time at the water's edge are discussed next in Chapter 6.

REFERENCES

Abel, D. C., Koenig, C. C. & Davis, W. P. (1987). Emersion in the mangrove forest fish, *Rivulus marmoratus*: a unique response to hydrogen sulfide. *Environmental Biology of Fishes* 18, 67–72.

Aguilar, N. M., Ishimatsu, A., Ogawa, K. & Huat, K. K. (2000). Aerial ventilatory responses of the mudskipper, *Periophthalmodon schlosseri*, to altered aerial and aquatic respiratory gas concentrations. *Comparative Biochemistry and Physiology* 127, 285–292.

Benson, K. R. (2002). The study of vertical zonation on rocky intertidal shores: a historical perspective. *Integrative and Comparative Biology* 42: 776–779.

Bhikajee, M. & Green, J. M. (2002). Behaviour and habitat of the Indian Ocean amphibious blenny, *Alticus monochrus*. *African Zoology* 37: 221–230.

Blaxter, J. H. S. & Batty, R. S. (1984). The herring swimbladder: loss and gain of gas. *Journal of the Marine Biology Association UK* 64, 441–459.

Brantley, R. K. & Bass, A. H. (1994). Alternative male spawning tactics and acoustic signals in the plainfin midshipman fish *Porichthys notatus* Girard (Teleostei, Batrachoididae). *Ethology* 96, 213–232.

Bridges, C. R. (1988). Respiratory adaptations in intertidal fish. *American Zoologist* 28, 79–96.

Brown, C. R., Gordon, M. S. & Martin, K. L. M. (1992). Aerial and aquatic oxygen uptake in the amphibious Red Sea rockskipper fish, *Alticus kirki* (family Blenniidae). *Copeia* 1992, 1007–1013.

Clayton, D. A. (1993). Mudskippers. *Oceanography and Marine Biology Annual Review* 31, 507–577.

Clayton, D. A. & Vaughan, T. C. (1986). Territorial acquisition in the mudskipper *Boleophthalmus boddarti* (Teleostei, Gobiidae) on the mud flats of Kuwait. *Journal of Zoology, London* 209A, 501–519.

Coleman, R. (1992). Reproductive biology and female parental care in the cockscomb prickleback, *Anoplarchus purpurescens* (Pisces: Stichaeidae). *Environmental Biology of Fishes* 41, 177–186.

Coleman, R. M. (1999). Parental care in intertidal fishes. In *Intertidal Fishes: Life in Two Worlds* (Horn, M. H., Martin, K. L. M. & Chotkowski, M. A., eds.), 165–180. San Diego: Academic Press.

Congleton, J. L. (1974). The respiratory response to asphyxia of *Typhlogobius californiensis* (Teleostei: Gobiidae) and some related gobies. *Biological Bulletin* 146, 186–205.

Congleton, J. L. (1980). Observations on the responses of some southern California tide pool fishes to nocturnal hypoxic stress. *Comparative Biochemistry and Physiology* 66A, 719–722.

Craig, P. M., Fitzpatrick, J. L., Walsh, P. J., Wood, C. M. & McClelland, G. B. (2014). Coping with aquatic hypoxia: how the plainfin midshipman (*Porichthys notatus*) tolerates the intertidal zone. *Environmental Biology of Fishes* 97, 163–172. DOI: 10.1007/s10641-013-0137-3.

Crane, J. (1981). Feeding and growth by the sessile larvae of the teleost *Porichthys notatus*. *Copeia* 1981, 895–897.

Dall, W. & Milward, N. E. (1969). Water intake, gut absorption and sodium fluxes in amphibious and aquatic fishes. *Comparative Biochemistry and Physiology* 30, 247–260.

Davenport, J. & Woolmington, A. D. (1981). Behavioural responses of some rocky shore fish exposed to adverse environmental conditions. *Marine Behavior and Physiology* 8, 1–12.

Daxboeck, C. & Heming, T. A. (1982). Bimodal respiration in the intertidal fish, *Xiphister atropurpureus* (Kittlitz). *Marine Behavior and Physiology* 9, 23–33.

Dejours, P. (1994). Environmental factors as determinants in bimodal breathing: an introductory overview. *American Zoologist* 34, 178–183.

Delaney, R. G., Lahiri, S. & Fishman, A. P. (1974). Aestivation of the African lungfish *Protopterus aethiopicus*: cardiovascular and respiratory functions. *Journal of Experimental Biology* 61, 111–128.

DeMartini, E. E. (1999). Intertidal spawning. In *Intertidal Fishes: Life in Two Worlds* (Horn, M. H., Martin, K. L. M. & Chotkowski, M. A., eds.), 143–164. San Diego: Academic Press.

Ebeling, A. W., Bernal, P. & Zuleta, A. (1970). Emersion of the amphibious Chilean clingfish, *Sicyases sanguineus*. *Biological Bulletin* 139, 115–137.

Edwards, D. G. & Cech, Jr., J. J. (1990). Aquatic and aerial metabolism of the juvenile monkeyface prickleback, *Cebidichthys violaceus*, an intertidal fish of California. *Comparative Biochemistry and Physiology* 96A, 61–65.

Eger, W. H. (1971). *Ecological and physiological adaptations of intertidal clingfishes (Teleostei: Gobeisocidae) in the northern Gulf of California.* Dissertation. University of Arizona.

Eiseley, L. (1946). *The Immense Journey.* New York: Random House, 211 pp.

Evans, D. H., Claiborne, J. B. & Kormanik, G. A. (1999). Osmoregulation, acid-base regulation, and nitrogen excretion. In *Intertidal Fishes: Life in Two Worlds* (Horn, M. H., Martin, K. L. M. & Chotkowski, M. A., eds.), 79–96. San Diego: Academic Press.

Feder, M. E. & Burggren, W. W. (1985). Cutaneous gas exchange in vertebrates: design, patterns, control and implications. *Biology Reviews* 60, 1–45.

Flostrand, L. A., Schweigert, J. F., Daniel, K. S. & Cleary, J. S. (2009). Measuring and modelling Pacific herring spawning-site fidelity and dispersal using tag-recovery dispersal curves. *ICES Journal of Marine Science* 66, 1754–1761.

Frank, K. T. & Leggett, W. C. (1981). Wind regulation of emergence times and early larval survival in capelin (*Mallotus villosus*). *Canadian Journal of Fisheries and Aquatic Sciences* 38, 215–223.

Garey, W. F. (1962). Cardiac responses of fishes in asphyxic environments. *Biological Bulletin* 122, 362–368.

Gee, J. H. & Gee, P. A. (1994). Aquatic surface respiration, buoyancy control and the evolution of air-breathing in gobies (Gobiidae: Pisces). *Journal of Experimental Biology* 198, 79–89.

Gonzales, T. T., Katoh, M., Ghaffar, M. A. & Ishimatsu, A. (2011). Gross and fine anatomy of the respiratory vasculature of the mudskipper *Periophthalmodon schlosseri* (Gobiidae: Oxudercinae). *Journal Morphology* 5, 629–640. DOI: 10.1002/jmor.10944.

Gordon, M. S., Fischer, S. & Terifeno, E. (1970). Aspects of the physiology of the terrestrial life in amphibious fishes. II. The Chilean clingfish *Sicyases sanguineus*. *Journal of Experimental Biology* 53, 559–572.

Gracey, A. Y., Lee, T.-H., Higashi, R. M. & Fan, T. (2011). Hypoxia-induced mobilization of stored triglycerides in the euryoxic goby *Gillichthys mirabilis*. *Journal of Experimental Biology* 214, 3005–3012.

Graham, J. B. (1976). Respiratory adaptations of marine air-breathing fishes. In *Respiration in Amphibious Vertebrates*, (Hughes, G. M., ed.), 165–187. London: Academic Press.

Graham, J. B. (1997). *Air-Breathing Fishes: Evolution, Diversity and Adaptation*. San Diego: Academic Press, 299 pp.

Graham, J. B. & Lee, H. J. (2004). Breathing air in air: in what ways might extant amphibious fish biology relate to prevailing concepts about early tetrapods, the evolution of vertebrate air-breathing, and the vertebrate land transition? *Physiological and Biochemical Zoology* 77, 720–731.

Graham, J. B. & Rosenblatt, R. H. (1970). Aerial vision: unique adaptation in an intertidal fish. *Science* 168, 386–388.

Graham, J. B., Jones, C. B. & Rubinoff, I. (1985). Behavioural, physiological, and ecological aspects of the amphibious life of the pearl blenny *Entomacrodus nigricans* gill. *Journal of Experimental Marine Biology and Ecology* 89, 255–268.

Graham, J. B., Rosenblatt, R. H. & Gans, C. (1978). Vertebrate air breathing arose in fresh waters and not in the oceans. *Evolution* 32, 459–463.

Halpin, P. M. & Martin, K. L. M. (1999). Aerial respiration in the salt marsh fish *Fundulus heteroclitus* (Fundulidae). *Copeia* 1999, 743–748.

Hart, J. L. (1973). *Pacific Fishes of Canada*. Fisheries Research Board of Canada Bulletin 180, 740 pp.

Hensley, D. A. & Courtenay, Jr., W. R. (1980). *Clarias batrachus* (Linnaeus) walking catfish. In *Atlas of North American Freshwater Fishes* (Lee, D. S., Gilbert, C. R., Hocutt, C. H., Jenkins, R. E., McAllister, D. E. & Stauffer, Jr., J. R., eds.), 475. North Carolina Biological Survey Publication #1980-12. North Carolina State Museum of Natural History.

Hill, J. V., Davison, W. & Marsden, I. D. (1996). Aspects of the respiratory biology of two New Zealand intertidal fishes, *Acanthoclinus fuscus* and *Forsterygion* sp. *Environmental Biology of Fishes* 45, 85–93.

Hillman, S. S. & Withers, P. C. (1987). Oxygen consumption during aerial activity in aquatic and amphibious fish. *Copeia* 1987, 232–234.

Hirose, T. & Kawaguchi, K. (1998). Spawning ecology of Japanese surf smelt, *Hypomesus pretiosus japonicas* (Osmeridae), in Otsuchi Bay, northeastern Japan. *Environmental Biology of Fishes* 52, 213–223.

Horn, M. H. & Gibson, R. N. (1988). Intertidal fishes. *Scientific American* 256, 64–70.

Horn, M. H. & Martin, K. L. (2006). Rocky intertidal zones. In *Ecology of Marine Fishes: California and Adjacent Waters* (Allen, L. A., Horn, M. H. & Pondella, D., eds.), 205–226. Berkeley: University of California Press.

Horn, M. H. & Riegle, K. C. (1981). Evaporative water loss and intertidal vertical distribution in relation to body size and morphology of stichaeoid fishes from California. *Journal of Experimental Marine Biology and Ecology* 50, 273–288.

Horn, M. H., Martin, K. L. M. & Chotkowski, M. A. (eds.). (1999). *Intertidal Fishes: Life in Two Worlds*. San Diego: Academic Press, 399 pp.

Hsieh, S.-T. T. (2010). A locomotor innovation enables water-land transition in a marine fish. *PLoS ONE* 5 (6), e11197. DOI:10.1371/journal.pone.0011197.

Huse, G. (1998). Sex-specific life history strategies in capelin (*Mallotus villosus*)? *Canadian Journal of Fisheries and Aquatic Science* 55, 631–638.

Innes, A. J. & Wells, R. M. G. (1985). Respiration and oxygen transport functions of the blood from an intertidal fish, *Helcogramma medium* (Tripterygiidae*). Environmental Biology of Fishes* 14, 213–226.

Ip, Y. K., Chew, S. F. & Tang, P. C. (1991). Evaporation and the turning behavior of the mudskipper, *Boleophthalmus boddaerti. Zoological Sciences* 8, 621–623.

Ip, Y. K., Chew, S. F. & Randall, D. J. (2002). Five tropical fishes, six different strategies to defend against ammonia toxicity on land. *Comparative Biochemistry and Physiology A* 134, S113–114.

Ishimatsu, A. (2012). Evolution of the cardiorespiratory system in air-breathing fishes. *Aquatic and Biological Science Monographs* 5, 1–28.

Ishimatsu, A. & Gonzales, T. T. (2011). Mudskippers: frontrunners in the modern invasion of land. In *The Biology of Gobies* (Patzner, R. A., Van Tassell, J. L., Kovacic, M. & Kapoor, B. G., eds.), 609–638. Enfield, NH: Taylor & Francis, CRC Press.

Ishimatsu, A. & Graham, J. G. (2011). Roles of environmental cues for embryonic incubation and hatching in mudskippers. *Integrative and Comparative Biology* 51, 38–48.

Ishimatsu, A., Takeda, T., Tsuhako, Y., Gonzales, T. T. & Khoo, K. H. (2009). Direct evidence for aerial egg deposition in the burrows of the Malaysian mudskipper, *Periophthalmodon schlosseri. Ichthyological Research* 56, 417–420.

Ishimatsu, A., Yoshida, Y., Itoki, N., Takeda, T., Lee, H. J. & Graham, J. B. (2007). Mudskippers brood their eggs in air but submerge them for hatching. *Journal Experimental Biology* 210, 3946–3954.

Iwama, G. K., Afonso, L. O., Todgham, A., Ackerman, P. & Nakano, K. (2004). Are hsps suitable for indicating stressed states in fish? *Journal of Experimental Biology* 207, 15–19.

Jew, C. J., Wegner, N. C., Yanagitsuru, Y., Tresguerres, M. & Graham, J. B. (2013). Atmospheric oxygen levels affect mudskipper terrestrial performance: implications for early tetrapods. *Integrative and Comparative Biology* 53, 248–257. DOI: 10.1093/icb/ict034.

Johansen, K. & Lenfant, C. (1968). Respiration in the African lungfish *Protopterus aethiopicus*. II. Control of breathing. *Journal of Experimental Biology* 49, 453–468.

Kok, W. K., Lim, C. B., Lam, T. J. & Ip, Y. K. (1998). The mudskipper *Periophthalmus scholosseri* respires more efficiently on land than in water and vice versa for *Boleophthalmus boddaerti. Journal of Experimental Zoology* 280, 86–90.

LeBlanc, D. M., Wood, C. M., Fudge, D. S. & Wright, P. A. (2010). A fish out of water: gill and skin remodeling promotes osmo- and ionoregulation in the mangrove killifish *Kryptolebius marmoratus. Physiological and Biochemical Zoology* 83, 932–949.

Low, W. P., Ip, Y. K. & Lane, D. J. W. (1990). A comparative study of the gill morphometry in the mudskippers—*Periophthalmus chrysospilos, Boleophthalmus boddaerti* and *Periophthalmodon schlosseri. Zoological Sciences* 7, 29–39.

Luck, A. & Martin, K. L. M. (1999). Tolerance of forced air emergence by a fish with a broad vertical distribution, the rockpool blenny, *Hypsoblennius gilberti* (Blenniidae). *Environmental Biology of Fishes* 54, 295–301.

Mandic, M., Sloman, K. A. & Richards, J. G. (2009a). Escaping to the surface: a phylogenetically independent analysis of hypoxia-induced respiratory behaviors in sculpins. *Physiological and Biochemical Zoology* 82, 703–738.

Mandic, M., Todgham, A. E. & Richards, J. G. (2009b). Mechanisms and evolution of hypoxia tolerance in fish. *Proceedings of the Royal Society B* 276, 735–744.

Marliave, J. B. & DeMartini, E. E. (1977). Parental behavior of intertidal fishes of the stichaeid genus *Xiphister*. *Canadian Journal of Zoology* 55, 60–63.

Martin, K. L. M. (1991). Facultative aerial respiration in an intertidal sculpin, *Clinocottus analis*. *Physiological Zoology* 64, 1341–1355.

Martin, K. L. M. (1993). Aerial release of CO_2 and respiratory exchange ratio in intertidal fishes out of water. *Environmental Biology of Fishes* 37, 189–196.

Martin, K. L. M. (1995). Time and tide wait for no fish: intertidal fishes out of water. *Environmental Biology of Fishes* 44, 165–181.

Martin, K. L. M. (1996). An ecological gradient in air-breathing ability among marine cottid fishes. *Physiological Zoology* 69, 1096–1113.

Martin, K. L. M. (1999). Ready and waiting: delayed hatching and extended incubation of anamniotic vertebrate terrestrial eggs. *American Zoologist* 39, 279–288.

Martin, K. L. M. (2013). Theme and variations: amphibious air-breathing intertidal fishes. *Journal of Fish Biology* 84, 577–602. DOI: 10.1111/jfb.12270.

Martin, K. L. M. & Bridges, C. R. (1999). Respiration in water and air. In *Intertidal Fishes: Life in Two Worlds* (Horn, M. H., Martin, K. L. M. & Chotkowski, M. A., eds.), 54–78. San Diego: Academic Press.

Martin, K. L. & Carter, A. L. (2013). Brave new propagules: terrestrial embryos in anamniotic eggs. *Integrative and Comparative Biology* 53, 233–247. DOI: 10.1093/icb/ict018.

Martin, K. L. M. & Lighton. J. R. B. (1989). Aerial CO_2 and O_2 exchange during terrestrial activity in an amphibious fish, *Alticus kirki* (Blenniidae). *Copeia* 1989, 723–727.

Martin, K. L. M. & Swiderski, D. L. (2001). Beach spawning in fishes: phylogenetic tests of hypotheses. *American Zoologist* 41, 526–537.

Martin, K. L., Bailey, K., Moravek, C. & Carlson, K. (2011). Taking the plunge: California grunion embryos emerge rapidly with environmentally cued hatching. *Integrative and Comparative Biology* 51, 26–37. DOI: 10.1093/icb/icr037.

Martin, K. L. M., Van Winkle, R. C., Drais, J. E. & Lakisic, H. (2004). Beach spawning fishes, terrestrial eggs, and air breathing. *Physiological and Biochemical Zoology* 77, 750–759.

Martinez, M. L., Landry, C., Boehm, R., Manning, S., Cheek, A. O. & Rees, B. B. (2006). Effects of long-term hypoxia on enzymes of carbohydrate metabolism in the Gulf killifish, *Fundulus grandis*. *Journal of Experimental Biology* 209, 3851–3861.

McBryan, T. L., Anttila, K., Healy, T. M. & Schulte, P. M. (2013). Responses to temperature and hypoxia as interacting stressors in fish: implications for adaptation to environmental change. *Integrative and Comparative Biology* 53, 648–659. DOI: 10.1093/icb/ict066.

Middaugh, D. (1981). Reproductive ecology and spawning periodicity of the Atlantic silverside, *Menidia menidia* (Pisces: Atherinidae). *Copeia* 1981, 766–776.

Middaugh, D. P., Kohl, H. W. & Burnett, L. E. (1983). Concurrent measurement of intertidal environmental variables and embryo survival for the California grunion, *Leuresthes tenuis*, and Atlantic silverside, *Menidia menidia* (Pisces: Atherinidae). *California Fish and Game* 69, 89–96.

Middaugh, D. P., Scott, G. I. & Dean, J. M. (1981). Reproductive behavior of the Atlantic silverside, *Menidia menidia* (Pisces: Atherinidae). *Environmental Biology of Fishes* 6, 269–276.

Nakano, K. & Iwama, G. K. (2002). The 70-kDa heat shock protein response in two intertidal sculpins, *Oligocottus maculosus* and *O. snyderi*: relationship of hsp 70 and thermal tolerance. *Comparative Biochemistry and Physiology A* 133, 79–94.

Nakashima, B. S. & Wheeler, J. P. (2002). Capelin (*Mallotus villosus*) spawning behavior in Newfoundland waters: the interaction between beach and demersal spawning. *ICES Journal of Marine Science* 59, 909–916.

Neider, J. (2005). Amphibious behaviour and feeding ecology of the four-eyed blenny (*Dialommus fuscus*, Labrisomidae) in the intertidal zone of the island of Santa Cruz (Galapagos, Ecuador). *Journal of Fish Biology* 58, 755–767. DOI: 10.1111/j.1095-8649.2001.tb00528.x.

Nilsson, G. E., Hobbs, J.-P. A., Ostlund-Nilsson, S. & Munday, P. L. (2007). Hypoxia tolerance and air-breathing ability correlate with habitat preference in coral-dwelling fishes. *Coral Reefs* 26, 241–248.

Pace, C. M. & Gibb, A. C. (2009). Mudskipper pectoral fin kinematics in aquatic and terrestrial environments. *Journal of Experimental Biology* 212, 2279–2286.

Pelster, B., Bridges, C. R. & Grieshaber, M. K. (1988). Physiological adaptations of the intertidal rockpool teleost *Blennius pholis* L., to aerial exposure. *Respiration Physiology* 71, 355–374.

Piiper, J. & Scheid, P. (1975). Gas transport efficacy of gills, lungs, and skin: theory and experimental data. *Respiration Physiology* 23, 209–221.

Ralston, S. L. & Horn, M. H. (1986). High tide movements of the temperate-zone herbivorous fish *Cebidichthys violaceus* (Girard) as determined by ultrasonic telemetry. *Journal of Experimental Marine Biology and Ecology* 98, 35–50.

Randall, D. J., Burggren, W. W., Farrell, A. P. & Haswell, M. S. (1981). *The Evolution of Air-Breathing in Vertebrates*. London: Cambridge University Press.

Randall, D. J., Ip, Y. K., Chew, S. F. & Wilson, J. M. (2004). Air breathing and ammonia excretion in the giant mudskipper, *Periophthalmus schlosseri*. *Physiological and Biochemical Zoology* 77, 783–788.

Richards, J. G. (2011). Physiological, behavioral and biochemical adaptations of intertidal fishes to hypoxia. *Journal of Experimental Biology* 214, 191–199. DOI: 10.1242/jeb.047951.

Richards, J. G., Sardella, B. A. & Schulte, P. M. (2008). Regulation of pyruvate dehydrogenase in the common killifish, *Fundulus heteroclitus*, during hypoxia exposure. *American Journal of Physiology* 295, R979–R990.

Riegle, K. C. (1976). *Oxygen Consumption, Heart Rates, Whole Body Lactate Levels, and Evaporative Water Loss in the Monkeyface Eel Cebidichthys violaceus (Family: Stichaeidae), an Amphibious Marine Fish from California*. MA thesis. California State University.

Russell, G. A., Middaugh, D. P. & Hemmer, M. J. (1987). Reproductive rhythmicity of the Atherinid fish, *Colpichthys regis*, from Estero Del Soldado, Sonora, Mexico. *California Fish and Game* 73, 169–174.

Saruwatari, T., Lopez, J. A., & Pietsch, T. W. (1997). A revision of the osmerid genus *Hypomesus* Gill (Teleostei: Salmoniformes), with the description of a new species from the southern Kuril Islands. *Species Diversity* 2, 59–82.

Sayer, M. D. J. & Davenport, J. (1991). Amphibious fish: why do they leave water? *Reviews in Fish Biology and Fisheries* 1, 159–181.

Scholander, P. F., Bradstreet, E. & Garey, W. F. (1962). Lactic acid response in the grunion. *Comparative Biochemistry and Physiology* 6, 201–203.

Shimizu, N., Sakai, Y., Hashimoto, H. & Gushima, K. (2006). Terrestrial reproduction by the air-breathing fish *Andamia tetradactyla* (Pisces: Blenniidae) on supralittoral reefs. *Journal of Zoology* 269, 357–364. DOI:10.1111/j.1469-7998.2006.00113.x.

Sloman, K. A., Mandic, M., Todgham, A. E., Fangue, N. A., Subrt, P. & Richards, J. G. (2008). The response of the tide pool sculpin, *Oligocottus maculosus*, to hypoxia in laboratory, mesocosm and field environments. *Comparative Biochemistry and Physiology* 149A, 284–292.

Sollid, J. & Nilsson, G. E. (2006). Plasticity of respiratory structures—adaptive remodeling of fish gills induced by ambient oxygen and temperature. *Respiratory Physiology and Neurobiology* 154, 241–251.

Spratt, J. D. (1986). The amazing grunion. *Marine Resource Leaflet No.*3, California Department of Fish and Game.

Stephenson, T. A. & Stephenson, A. (1949). The universal features of zonation between tide-marks on rocky coasts. *Journal of Ecology* 37, 289–305.

Strathmann, R. R. & Hess, H. C. (1999). Two designs of marine egg masses and their divergent consequences for oxygen supply and desiccation in air. *American Zoologist* 39, 253–260.

Sweetnam, D. A., Baxter, R. D. & Moyle, P. B. (2001). True smelts. In *California's Living Marine Resources: A Status Report*, 470–478. Sacramento: California Department of Fish and Game.

Swenson, R. O. (1997). Sex-role reversal in the tidewater goby, *Eucyclogobius newberryi*. *Environmental Biology of Fishes* 50, 27–40. DOI:10.1023/A:1007352704614.

Tamura, S. O., Morii, H. & Yuzuriha, M. (1976). Respiration of the amphibious fishes *Periophthalmus cantonensis* and *Boleophthalmus chinensis* in water and on land. *Journal of Experimental Biology* 65, 97–107.

Taylor, M. (1984). Lunar synchronization of fish reproduction. *Transactions of the American Fisheries Society* 113: 484–495.

Taylor, M. H., DiMichele, L. & Leach, G. J. (1977). Egg stranding in the life cycle of the mummichog, *Fundulus heteroclitus*. *Copeia* 1977, 397–399.

Teal, J. M. & Carey, F. G. (1967). Skin respiration and oxygen debt in the mudskipper *Periophthalmus sobrinus*. *Copeia* 1967, 677–679.

Templeman, W. (1948). The life history of the capelin (*Mallotus villosus*) in Newfoundland waters. *Bulletin of the Newfoundland Government Laboratory at St. John's* 17, 1–155.

Tewksbury, H. T. & Conover, D. O. (1987). Adaptive significance of intertidal egg deposition in the Atlantic silverside *Menidia menidia*. *Copeia* 1987, 76–83.

Thomson, D. A. & Muench, K. A. (1976). Influence of tides and waves on the spawning behavior of the Gulf of California grunion, *Leuresthes sardina* (Jenkins and Evermann). *Bulletin of the Southern California Academy of Sciences* 75, 198–203.

Todd, E. S. (1968). Terrestrial sojourns of the longjaw mudsucker, *Gillichthys mirabilis*. *Copeia* 1968, 192–194.

Todd, E. S. (1973). Positive buoyancy and air-breathing: a new piscine gas bladder function. *Copeia* 1973, 461–464.

Todd, E. S. & Ebeling, A. W. (1966). Aerial respiration in the longjaw mudsucker *Gillichthys mirabilis* (Teleostei: Godiidae). *Biological Bulletin* 130, 265–288.

Todgham, A. E., Schulte, P. M. & Iwama, G. K. (2005). Cross-tolerance in the tide pool sculpin: the role of heat shock proteins. *Physiological and Biochemical Zoology* 78, 133–144. http://www.jstor.org/stable/10.1086/425205.

Truchot, J. P. & Duhamel-Jouve, A. (1980). Oxygen and carbon dioxide in the marine inter-tidal environment: diurnal and tidal changes in rockpools. *Respiration Physiology* 39, 241–254.

Tzaneva, V., Bailey, S. & Perry, S. F. (2011). The interactive effects of hypoxemia, hyper-oxia, and temperature on the gill morphology of goldfish (*Carassius auratus*). *American Journal of Physiology* 300, R1344–R1351.

Uchiyama, M., Komiyama, M., Yoshizawa, H., Shimizu, N., Konno, N., & Matsuda, K. (2012). Structures and immunolocalization of Na+, K+ -ATPase, Na+ /H+ exchanger 3 and vacuolar-type H+ -ATPase in the gills of blennies (*Teleostei: Blenniidae*) inhabiting rocky intertidal areas. *Journal of Fish Biology* 80, 2236–2252. DOI: 10.111/j.1095-8649.2012.

Vasconcelos, R. O., Sarrico, R., Ramos, A., Modesto, T., Fonseca, P. J., & Amorin, M. C. P. (2011). Vocal behavior predicts reproductive success in a teleost fish. *Behavioral Ecology* 23, 375–383. DOI: 10.1093/beheco/arr199.

Walker, B. (1949). *Periodicity of spawning by the grunion, Leuresthes tenuis, an Atherine fish*. PhD Dissertation. University of California, Los Angeles; Scripps Institution of Oceanography. http://www.escholarship.org/uc/item/1j33928x.

Walker, B. (1952). A guide to the grunion. *California Fish and Game* 38, 409–420.

Watters, J. V. & Cech, Jr., J. J. (2003). Behavioral responses of mosshead and woolly sculpins to increasing environmental hypoxia. *Copeia* 2003, 397–401.

Wilson, B., Batty, R. S., & Dill, L. M. (2004). Pacific and Atlantic herring produce burst pulse sounds. In *Proceedings of the Royal Society of London B* 271, S95–S97. DOI: 10.1098/rsbl.2003.0107.

Wright, P. A. (2012). Environmental physiology of the mangrove rivulus, *Kryptolebias marmoratus*, a cutaneously breathing fish that survives for weeks out of water. *Integrative and Comparative Biology* 52, 792–800. DOI:10.1093/icb/ics091.

Wright, W. G. & Raymond, J. A. (1978). Air-breathing in a California sculpin. *Journal of Experimental Zoology* 203, 171–176.

Yadav, A. N. & Singh, B. R. (1989). Gross structure and dimensions of the gill in an air-breathing estuarine goby, *Pseudapocryptes lanceolatus*. *Japanese Journal of Ichthyology* 36, 252–259.

Yamahira, K. (1996). The role of intertidal egg deposition on survival of the puffer, *Takifugu niphobles* (Tetraodontidae). *Environmental Biology of Fishes* 40, 255–261.

Yoshiyama, R. M. & Cech, Jr., J. J. (1994). Aerial respiration by rocky intertidal fishes of California and Oregon. *Copeia* 1994, 153–158.

Yoshiyama, R. M., Valey, C. J., Schalk, L. L., Oswald, N. M., Vaness, K. K., Lauritzen, D. & Limm, M. (1995). Differential propensities for aerial emergence in intertidal sculpins (Teleostei: Cottidae). *Journal of Experimental Marine Biology and Ecology* 191, 195–207.

Zander, C. D. (1972). Beziehungen zwischen Korperbau und Lebensweise bei Blenniidae (Pisces) aus dem Roten Meer. I. Aussere Morphologie. *Marine Biology* 13, 238–246.

Zander, C. D. (1983). Terrestrial sojourns of two Mediterranean Blennioid fish (Pisces, Blenniodei). *Senckenbergiana Maritime* 15, 19–26.

Zander, C. D., Nieder, J. & Martin, K. L. M. (1999). Vertical distribution patterns. In *Intertidal Fishes: Life in Two Worlds* (Horn, M. H., Martin, K. L. M. & Chotkowski, M. A., eds.), 26–53. San Diego: Academic Press.

Zhang, Q.-Y., Taniguchi, T., Takita, T. & Ali, A. B. (2000). On the epidermal structure of *Boleophthalmus* and *Scartelaos* mudskippers with reference to their adaptation to terrestrial life. *Ichthyological Research* 47, 359–366.

Zheng, W., Wang, Z., Collins, J. E., Andrews, R. M., Stemple, D. & Gong, Z. (2011). Comparative transcriptome analyses indicate molecular homology of zebrafish swimbladder and mammalian lung. *PLoS One* 6: e24019. DOI: 10.1371/journal.pone.0024019.

6 Unsafe Sex: The Dangers of Novel Predators and Other Terrestrial Influences for Beach-Spawning Fishes and Their Embryos

One frequently suggested explanation for the evolution of beach spawning in fishes is that this helps them avoid predation. This is appealing, particularly when considering those species that only spawn under cover of darkness. However, this explanation does not hold up particularly well to scrutiny. Large aggregations of active fishes that occur predictably with the tides in preparation for spawning are also a regularly scheduled temptation for many predators, including fishes, invertebrates, birds, and mammals.

This chapter considers the various marine and terrestrial animals that feast on beach-spawning fishes and their propagules. Predation on fishes during spawning may slow or stop a run. Predation on nests and embryos during incubation on beaches may cause significant losses. Nest guarding to avoid those losses of eggs may threaten dutiful parents. Opportunistic fisheries have arisen on various species of beach-spawning fishes, making humans predators on beach-spawning runs as well.

Predatory behavior is exciting and interesting to watch. It is unfortunate that there are few studies of predatory behavior and its consequences during spawning aggregations and runs of beach-spawning fishes. As might be predicted, most observations have been made on those species of beach-spawning fishes with the most extreme or obvious spawning behavior or those that produce the largest aggregations. Fortunately, runs of the capelin *Mallotus villosus* (Osmeridae) and two silversides, the California grunion *Leuresthes tenuis* and the Atlantic silverside *Menidia menidia*, are frequently attended by observant scientists, who have noticed many different predators.

6.1 MARINE AND AVIAN PREDATORS ATTACK BEACH-SPAWNING FISHES IN THE OCEAN

Forage fishes are important links in the marine food web and their depletion can have serious consequences for top predators (Pauly et al., 1998; Coll et al., 2006; Kaschner et al., 2006). Many seabirds and marine mammals eat small fishes, taking advantage of their schools at sea. Routine predation on fishes happens all the time, but when a species comes together in preparation for spawning, the larger numbers and greater activity may attract more attention.

Large numbers of fishes in a school or a spawning aggregation attract multiple predators (Vermeer & Devito, 1986; Sancho, 2000; Sancho et al., 2000), but with few exceptions, during spawning in the pelagic realm, actual take and kills of adult fish are rare (Moyer, 1987). This is not the case for beach-spawning fishes, in part because of the shallow water and in part because one major avenue of escape is blocked by the shoreline.

Some subtidal fishes that migrate in to spawn on beaches form large temporary aggregations near shore before starting to spawn. Huge schools of capelin *Mallotus villosus* that converge along the coast of Canada provide food for mammals including humpback whales and migratory harp seals, birds such as black-legged kittiwakes, northern gannets, and puffins during their nesting season. The economically important fisheries for Atlantic cod and haddock depend on *M. villosus* for sustenance.

When California grunion *Leuresthes tenuis* mass near shore during the day before a spawning run, it is common to see brown pelicans *Pelicanus occidentalis*, common dolphins *Delphinus delphis*, and California sea lions *Zalophus californiensis* enjoying a feast just beyond the breaking surf. Even an invertebrate predator, the Humboldt squid *Dosidicus gigas*, chases *L. tenuis* in great schools (Martin et al., 2011). The importance of the California grunion *L. tenuis* in the marine food web is indicated by the many other predators that consume them at sea including mammals such as harbor seals *Phoca vitulina*; birds including least terns *Sterna antillarum*, elegant terns *Thalasseus elegans*, and black skimmers *Rynchops niger*; and fishes such as queenfish *Seriphus politus*, California halibut *Paralichthys californicus*, corbina *Menticirrhus undulatus*, leopard sharks *Triakis semifasciata*, great white sharks *Carcharodon carcharias*, and shovel-nose guitarfish *Rhinobatos productus* (Gregory, 2001; Elliott et al., 2007; Sandrozinsky, 2013).

6.2 MARINE AND TERRESTRIAL PREDATORS ATTACK DURING BEACH-SPAWNING RUNS

Spawning runs that occur predictably with the tides may have predators on shore waiting for them, as well as marine predators that pressure them from the sea. Some marine predators are so focused on following their prey that they may beach themselves during the pursuit. During spawning runs, marine predators may follow the California grunion *L. tenuis* up onto beaches in search of a meal (Martin et al., 2011). Small gray smoothhound sharks *Mustelus californicus*, leopard sharks *T. semifasciata*, and the shovel-nosed guitarfish *Rhinobatos productus* have been observed following California grunion up onto shore in the surf during runs, occasionally even fully emerging from the waves out of water (Grunion Greeters, personal communications). A receding wave may deposit a young sea lion on the beach during a run of *L. tenuis* to rest, satiated after feeding. Mass strandings of the Humboldt squid *Dosidocus gigas* occur when they follow grunion in the waves and fail to realize that they lack the grunion's knowledge of how to return to sea (Martin et al., 2011).

In addition to predators from the marine environment, *L. tenuis* become vulnerable to novel terrestrial predators during their spawning runs onto beaches. In some urban areas, domestic dogs and feral cats prey on runs of *L. tenuis*. Terrestrial predators include wildlife such as raccoons, rats, and coyotes that target the runs. Many shorebirds such as great blue herons *Ardea herodia*, black-crowned night herons *Nycticorax nycticorax*, snowy egrets *Egretta thula*, and even western gulls *Larus occidentalis* gobble up *L. tenuis* during runs until they become so heavy they are unable to fly (Figure 6.1). On some beaches, herons line up in anticipation of the feast even before the fish begin to appear on shore (Griem & Martin, 1997; Martin & Raim, 2014).

The tides that are conducive to spawning runs in *L. tenuis* occur only at night during the breeding season. As a result, these birds are up late on these nights, but not present on shore on nights when no runs may be expected. Beaches close to nesting or roosting areas for shorebirds are more likely to have nocturnal avian predators than beaches but less habitat for birds, even if both have equally strong spawning runs of *L. tenuis* (Martin & Raim, 2014). These shorebirds are not able to hunt for *L. tenuis* at any time other than during the spawning runs.

FIGURE 6.1   Black-crowned night heron *Nyctocorax nyctocorax* and great blue heron *Ardea herodia* await California grunion *Leuresthes tenuis* at the nighttime spawning run. (Photos by Peter Heistand (top) and Joe McLain (bottom). With permission.)

Barn owls *Tyto alba* have been reported chasing *L. tenuis* on beaches (Gallup, 1949). Fossil fish remains deposited by owls have been found in caves (Broughton et al., 2006). However, owl predation on grunion runs has not been observed since the 1940s even though hundreds of runs of *L. tenuis* have been observed over the past 10 years and *T. alba* still occupies some areas of the coast of California near beaches where *L. tenuis* runs occur.

Most birds are visual predators, so foraging depends on available light and is usually diurnal, with some exceptions. Artificial lights may increase vulnerability of spawning fishes to nocturnal predators (Rich & Longcore, 2006; Santos et al., 2010). In Southern California, the habitat of *L. tenuis,* many coastal locations have bright lights from nearby streets, businesses, piers, homes, and other structures that may aid predators in locating and more efficiently capturing this ephemeral prey resource during spawning runs. Predatory birds are present on beaches waiting for the spawning runs of *L. tenuis* whether the moon is full, new, or hidden by overcast skies (Martin & Raim, 2014). Black-crowned night heron eyes are specialized for night vision (Katzir & Martin, 1998), and they have been observed moving into pools of light from shore flashlights, when hunting *L. tenuis* on dark shores.

L. tenuis always spawns nocturnally. Its congener, the Gulf grunion *L. sardina,* spawns sometimes in the daytime because the tides are on a different schedule within the Gulf of California (see Chapter 4). Birds prey heavily on Gulf grunion during these March and April runs, including the brown pelican *Pelicanus occidentalis,* a cormorant *Phalocrocorax* sp., and several species of seagulls, *Larus occidentalis, L. delawarensis, L. californicus, L. heermani,* and *L. atricilla* (Thomson & Muench, 1976). These diurnal avian predators differ from those shorebirds that wade into spear fish during spawning runs of *L. tenuis,* although brown pelicans and cormorants prey upon the offshore aggregations of *L. tenuis* that form during the days before the nocturnal runs. *L. sardina* spawn mostly at the water's edge rather than out of water like *L. tenuis,* so these birds are feeding on the fish in shallow water rather than on shore (Figure 6.2).

Beach-spawning runs of *M. villosus* are not regularly predictable with the tides (Carscadden et al., 1989), so terrestrial predators on capelin runs are likely

FIGURE 6.2 Birds chasing Gulf grunion *Leuresthes sardina* during a daytime spawning run. (Photo by Tim Bourke. With permission.)

(a)

(b)

FIGURE 2.1 Vertical height in the intertidal zone is correlated with duration of air emergence. Organisms found higher in the intertidal zone will be emerged for longer periods of time and more frequently than organisms found lower in the intertidal zone. (a) Big Rock Beach at high tide. (b) The same beach at low tide.

FIGURE 4.2 Capelin *Mallotus villosus* triplet rolling on a Canadian beach. Many eggs can be seen already on the gravel. (Photo by Anna Olafsdottir.)

(a)

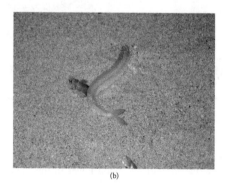

(b)

FIGURE 4.3 (a) A spawning run of California grunion *Leuresthes tenuis* out of water on a sandy beach. (Photo by Bill Hootkins.) (b) The female digs in vertically while the male fertilizes the eggs from the sand surface. (Photo by Doug Martin.)

FIGURE 6.4 Recreational fishing is a family activity for *Mallotus villosus*, the capelin roll, on a beach in Newfoundland, Canada. (Photo by Kenneth Frank.)

FIGURE 6.5 Recent hatchlings of the whitebait *Galaxias aculeatus* are the basis of this unique sandwich, served fresh seasonally in New Zealand. (Photo by Karen Martin.)

FIGURE 9.3 Recreational anglers on shore take fish during a spawning run of Gulf grunion. (Photo by Tim Bourque. With permission.)

FIGURE 10.2 Grunion Greeters find their target species on California beaches. (Photo by Jennifer Flannery Harr. With permission.)

FIGURE 10.5 The spawning run of the kusafugu puffer. (Photo used by permission of Hikari City Board of Education, Culture, and Lifelong Learning Division, http://www.city. hikari.lg.jp/bunka/kusahugu25.html.)

to be opportunistic. Because many *M. villosus* die after spawning, carrion feeders including bears and bald eagles may find their next meal on the shoreline after the run, feeding on the spent fish carcasses.

Daytime spawning aggregations of the beach-spawning Atlantic silverside fish *Menidia menidia* occur during high tides over sea grass beds (Middaugh, 1981). Many birds are predators on these aggregations, including the brown pelican *Pelicanus occidentalis carolinensis*, the common egret *Casmerodius album egretta*, the snowy egret *Egretta thula*, the least tern *Sterna albifrons*, the common tern *S. hirundo*, Forster's tern *S. forsteri*, the laughing gull *Larus atricilla*, the cormorant *Phalacrocorax auritus*, and the black skimmer *Rhynchops nigra nigra*. Fishes that prey upon spawning *M. menidia* include the spotted seatrout *Cynoscion nebulosus*, the bluefish *Pomatomus saltatrix*, the longnose gar *Lepisosteus osseus*, the Atlantic stingray *Dasyatis sabina*, and the sandbar shark *Carcharhinus plumbeus*. The diamondback terrapin *Malaclemys terrapin* also feeds on *M. menidia*. Avian and marine predators may substantially deplete the numbers of fishes in spawning aggregations of *M. menidia* (Takita et al., 1984).

Pacific herring *Clupea pallasii pallasii* run in huge numbers in San Francisco Bay and other estuaries during winter. The runs over eelgrass are associated with aggregations of seabirds such as common murres (Piatt & Anderson, 1996) and other predators, including the northern fur seal *Callorhinus ursinus*, sea lions *Zalophus californiensis*, killer whales *Orcinus orca*, dogfish *Mustelis canis*, and salmon *Oncorhynchus* (Hourston & Haegle, 1980). Offshore when they migrate into the pelagic realm, they are also eaten by hake *Merluccius productus*, sablefish *Anoplopoma fimbria*, and Pacific cod *Gadus macrocephalus* (Perez & Bigg, 1986; Lassuy & Moran, 1989).

6.3 PREDATION OCCURS ON INCUBATING EGGS AND EMBRYOS ON BEACHES

Horseshoe crabs *Limulus polyphemus* spawn tidally on beaches on the Atlantic coast of North America, where their eggs are important food resources for migrating birds such as red knots *Calidryas canutus rufus* (McGowan et al., 2011). Recognition of the role of these beach nursery areas in the food web, particularly for endangered bird species, has stimulated conservation efforts for this beach-spawning invertebrate, an unusual feat of public persuasion (Smith et al., 2011).

Similarly, the eggs of many beach-spawning fishes are consumed by many different species of shorebirds and other small predators, including invertebrates. Eggs of *L. tenuis* are buried beneath about 10 centimeters of sand on the California coast (Walker, 1952). Despite being hidden, the eggs can be found and preyed upon by many shorebirds including migrating birds such as the horned lark *Eremophila alpestris* (Olson, 1950), ruddy turnstones *Arenaria interpres*, long-billed curlews *Numenius americanus*, marbled godwits *Limosa fedoa*, willets *Tringa semipalmata*, sanderlings *Calidris alba*, western snowy plovers *Charadrius alexandrinus*, and other shorebirds (Walker, 1949; MacGinitie & MacGinitie, 1949; Middaugh et al., 1983). Urban birds such as pigeons *Columba livia* and American crows *Corvus brachyrhynchos* prey on nests of California grunion (Martin et al., 2011).

Birds that usually probe for sand-dwelling invertebrates may come across grunion eggs purely by chance. When this happens, a single individual can take advantage and completely devour thousands of eggs, the entire clutch of one female, in a few minutes (personal observation). A flock of birds that finds a patch of grunion nests can devastate the cohort in short order. Typically this type of predation occurs within a few hours after the run occurred, during the early daylight hours, and it may result from visual evidence of spawning activity on the beach such as holes where the fish dug in, or exposed eggs that were accidentally ejected from nests as females made tail flips while exiting. These visual signs fade from view after exposure to a few hours of sun and wind, and predation by birds on these clutches ceases until after the next run.

Along with the birds, small mammals such as ground squirrels *Otospermophilus beecheyi* (Olson, 1950; Figure 6.3) and raccoons *Procyon lotor* (personal observation) have learned to dig up the clutches of *L. tenuis* on the morning after a run, and can quickly consume a large number of eggs for breakfast from multiple clutches. Like bird predation, mammal predation seems to occur only on the days directly following a spawning run, and clutches that are not found during the first day or so will not be dug up later during incubation.

The eggs of *Menidia menidia* that were experimentally placed in cages at different tidal heights showed differences in survival. Placing the eggs uncaged in the subtidal zone exposed them to aquatic predators such as *Fundulus heteroclitus,* resulting in very low survival (Tewksbury & Conover, 1987). Blue crabs *Callinectes sapidus* prey on *Menidia* eggs in the intertidal zone (Middaugh, 1981). Birds such as the ruddy turnstone *Arenaria interpres* and the semipalmated sandpiper *Calidris pusilla* feed on *M. menidia* eggs, and Middaugh (1981) reported that two intact embryos were still alive after passing through the digestive tract of a ruddy turnstone.

FIGURE 6.3 Ground squirrel *Otospermophilus beecheyi* digging up eggs of California grunion *Leuresthes tenuis* the morning following a late night run. (Photo by Karen Martin.)

Eggs of *Leuresthes tenuis* that escape predation from birds and mammals may still be continuously exposed to invertebrate predators that lurk in the interstitial spaces of the sand. Sandworm polychaetes *Glycera americanus* lurk beneath the surface and feed on decaying or fresh organic matter. They are only too happy to nosh on grunion eggs. An old fisherman once advised the author that the best bait for catching halibut was these sandworms after they fed on grunion eggs. On some beaches used by *L. tenuis* for spawning, large numbers of sandworms are present. This may severely limit the success of clutches there, as one worm can completely consume an entire clutch of eggs over the incubation period. Insects including beetles and kelp flies do not live in the sand, but they lay their eggs on kelp and other organic matter. Once the eggs hatch, the voracious larvae feed on whatever they find. When meals are scarce and far between, these larvae capitalize on grunion nests if they can find them.

Pacific herring *Clupea p. pallasii* spawn on intertidal sea grass beds. Herring eggs deeper in the subtidal zone have decreased survival, presumably because of lower oxygen availability (Taylor, 1971), while placing the eggs higher in the intertidal zone results in relatively low losses from invertebrates and seabirds (Haegle 1993a,b). Seabirds known to eat herring eggs in California include the California gull *Larus californicus*, mew gull *L. canus*, Glaucous-winged gull *L. glaucescens*, western gull *L. occidentalis*, American coot *Fulica americana*, and surf scoter *Melanitta perspicillata* (Hardwick, 1973). Humans are also important egg predators for herring; see Section 6.6.

6.4   EGG PREDATION OCCURS BY CANNIBALISM DURING SPAWNING AND INCUBATION ON THE BEACH

Conspecific predation, or cannibalism, is a fact of life for many fishes, including beach-spawning fish species. Reports of adult fishes eating eggs or larvae of their own species are found in the scientific literature with regularity, no matter what the habitat, even if those eggs are their own offspring (Manica, 2002). Egg consumption may occur as part of nest guarding, to clean out decrepit or diseased eggs, or shortly after spawning, with adults feeding on freshly produced embryos. Cannibalism may affect both present and future reproductive success if the eggs are an alternative food source for a guarding parent (Manica, 2002).

Cannibalism of conspecific eggs occurs opportunistically by *Menidia menidia* (Kendall, 1902, cited by Middaugh, 1981). Pacific herring *Clupea p. pallasii* consume conspecific eggs and larvae "voraciously" (Hourston & Haegle, 1980). The mummichog *Fundulus heteroclitus* consumes both its own eggs and those of *M. menidia* in estuaries where they co-occur, both at the time of spawning and throughout incubation (Conover & Kynard, 1984; Able & Hata, 1984; Tewksbury & Conover, 1987). Sticklebacks *Gasterosseus aculeatus*, those textbook parents, also cannibalize if the opportunity arises (FitzGerald & van Havre, 1987). To kick it up a notch, females can distinguish their own eggs from those of others. A female stickleback may attack a nest and destroy its eggs to prepare it to receive her own clutch, suggesting an extreme response to competition for the services of the male guarding parent (FitzGerald & van Havre, 1987).

A female California grunion *Leuresthes tenuis* usually hides her eggs by burying them under sand. After she flips her tail out of the hole to return to the ocean, a few eggs may catch, follow, and turn up on the surface of sand. Those eggs may be eaten by another *L. tenuis* in the spawning run (Cavanagh et al., 2014).

Some beach-spawning fishes produce toxins to protect their eggs from predation. The eggs of the kusafugu puffer *Takifugu niphobles*, like those of other puffers, are toxic and unpalatable (Gladstone, 1987). One would think that avoidance of predation should not be a selection pressure for beach spawning in this species. However, cannibalism on eggs of *T. niphobles* has been observed during spawning (Yamahira, 1996). Perhaps placing the embryos at the water's edge limits the access of conspecifics to these propagules to just the time around the highest tides.

Similarly, many intertidal species of sculpins have brightly colored toxic eggs (Pillsbury, 1957). These species generally do not guard their eggs or nests. Cannibalism is well known among freshwater and subtidal sculpin species (DeMartini, 1976; Allen & Horn, 2006) and has been reported for sculpins in the genus *Artedius* (Petersen et al., 2005) although not yet for other intertidal sculpins (Pfister, 2007).

6.5 PREDATION MAY OCCUR IN THE NEST ON GUARDING PARENTS AND EMBRYOS

Among the fishes that spawn in the rocky intertidal zone, nest guarding is seen in several families including Blenniidae, Gobiidae, Xiphisteriidae, Batrachoididae, Cottidae, and Stichaeidae. (See Chapters 3 and 4 for more on parental care.) Parental attention reduces nest predation, but this may be less critical in the intertidal zone than in the subtidal zone (Goncalves & Almada, 1998). However, when guarding parents are removed, egg survival decreases.

Although the predators of eggs typically are different from the predators of adult fishes, it is reasonable to expect that there is a cost to the parent for remaining with the eggs, even if there is seemingly little overt activity near the nest, and even if the fishes are resident in the intertidal zone throughout the year. Part of this cost may be increased risk of predation because of reluctance to move away from the nest, and there may also be an energetic cost related to decreased foraging time when guarding the clutch of eggs. The male *Porichthys notatus* that stay guarding multiple clutches in the intertidal zone, even after the hatchlings have emerged, look emaciated and injured as time progresses, and the hypothesis is that mortality is high for these guarding parents (Feder et al., 1974).

It is likely that terrestrial mammalian, avian, and reptilian predators opportunistically feed on intertidal fishes while they are guarding nests. Unfortunately little data exist regarding predatory risks to guarding parents for intertidal nests. Many nest-guarding species place the eggs within a chamber created by space under rocks or in crevices, or within shells. This controls the exposure of the eggs and the adult but may also make escape more difficult if the location is discovered by predators.

Males of the intertidally spawning sand goby *Pomatoschistus minutus* guard nests; excluding avian predators results in a larger size of adult males at those protected nests after a week than at unprotected nests, suggesting that predators selectively removed larger males while they were vulnerable (Lindstrom & Ranta, 1992). During the nesting season it is not difficult to find guarding adults of *Porichthys notatus*, *Anoplarchus purpurescens*, or *Xiphister mucosus* by turning over intertidal rocks and boulders, and they are easy prey for human collectors. It is possible that predation on adult fishes is cryptic and may be more prevalent than has been reported.

Snakes have been reported to feed on intertidal fishes. The garter snake *Thamnophis elegans* eats *Anoplarchus purpurescens* on rocky beaches in Friday Harbor, Washington, and in British Columbia, Canada (Batts, 1961; Gregory, 1978), in an area where *A. purpurescens* is known to guard its nests (Coleman, 1992). Cottonmouths *Agkistrodon piscivorus* feed on dropped fish and dead carrion that wash into intertidal zone in Florida, and sometimes these snakes enter seawater (Lillywhite et al., 2008). This fondness for fish dinners has been suggested as a possible route to secondary marine invasion for terrestrial snakes. A turtle, the diamondback terrapin, eats the beach-spawning fish *Menidia menidia* (Middaugh, 1981). It is reasonable to hypothesize that other reptiles may prey upon the small forage fishes that spawn on beaches.

Beaches seem to inspire novel predatory behaviors. Local groups of aquatic animals emerge to feed on beaches in several locations. Dolphins *Tursiops truncatus* in South Carolina herd aquatic fishes up onto shore to make it easy for these predators to capture and eat their helpless prey (Duffy-Echevarria et al., 2008). Orcas *Orcinus orca* in Patagonia surf onto shore in waves to prey upon fur seals hauled out on beaches (Lopez & Lopez, 1985). A recent report described the successful capture by catfishes *Siluris glanis* of pigeons, *Columbia livia*, on a freshwater beach in France (Cucherousset et al., 2012). It is appealing to consider the possibility (albeit remote) that this kind of behavior by beach-spawning fishes could turn the tables on their attackers. Predatory behaviors appear to be as malleable as spawning behaviors on beaches. This is an area that deserves additional study.

6.6 HUMAN FISHERIES TARGET SOME BEACH-SPAWNING FISHES

Humans can recognize the signs of impending spawning runs as easily as animals can, and artisanal fisheries have arisen on beach-spawning runs in many different locales. The coastal setting in shallow water allows individuals, alone or in small groups, to harvest successfully with simple equipment. The ease of capture of the distracted fishes, and their spawning sites near large human populations, place some species at risk of overfishing.

Pacific herring *Clupea p. pallasii* were important for First Nations and Native Americans as food. Native Alaskans also harvest the spawn on kelp and surfgrass as a traditional food (Hourston & Haegle, 1980).

Herring have also supported commercial fishing for over a century (Lassuy & Moran, 1989). Initially caught as adults and salted, canned, or used as bait, they were also used in the reduction industry for fertilizer and animal feed. Since the 1970s, a new fishery targeted herring roe for the Japanese market (Lassuy & Moran, 1989). Most of this harvest is exported from North America to Japan, with a small fraction sold in North America. In the 1980s some harvest of spawn-on-kelp began as a high-end contribution to the market for fine-dining restaurants and for gift giving.

Herring adults may be impounded briefly in nets as they mass before a spawning run and provided with kelp as a spawning substrate to improve production. Thicker layers of eggs coating the vegetation command higher prices. This specialty fishery takes place primarily along the Pacific coast of British Columbia and Alaska.

Smelts (Osmeridae) are small, silver fishes that may be present in huge numbers. In some areas, surf smelt *Hypomesus pretiosus* and night smelt *Spirinchus starksi* have recreational fisheries involving cast nets or fish rakes. Surf smelt are occasionally caught for bait, but their small size gives them little commercial importance (Sweetnam et al., 2001). The current status of these fisheries is not clear. A similar anadromous smelt, the eulachon *Thaleichthys pacificus*, is called the candlefish because it is so full of oil that it can burn with a bright flame. It was once so plentiful that it was fished out of streams by the thousands by Native Americans and others. Presently it is listed as a federally endangered species and rarely seen at all (Duran, 2008).

Another smelt, the capelin *Mallotus villosus*, makes spawning runs in Newfoundland, Canada. Families view the capelin roll as an adventurous way to get an unusual snack rather than as a source of sustenance. Neighbors alert neighbors as soon as this unpredictable event begins. Capelin are fished commercially out at sea, but the fishery on the beach during the spawning run is strictly recreational (Figure 6.4).

Capelin are fished commercially in the Atlantic, in Europe, and in Iceland but not in the Pacific Ocean (Gjøsæter, 1998). The population fluctuates dramatically with dynamics that are not well understood, and thus setting catch limits is difficult. In the past, capelin were harvested for bait, food, and dog food as well as use as fertilizer in Canada, Russia, and other northern nations (Carscadden, 1981). Today the main commercial use of *M. villosus* is as fish meal for processed foods, and they are also fed to captive toothed whales and dolphins in oceanariums. See Chapter 9 for more on this fishery.

Unlike the opportunistic recreational fishing that takes place on remote rural beaches in Newfoundland, runs of California grunion *L. tenuis* are all too easily predictable, down to the hour, because of tidal synchronization. Human predation is

FIGURE 6.4 (See color insert.) Recreational fishing is a family activity for *Mallotus villosus*, the capelin roll, on a beach in Newfoundland, Canada. (Photo by Kenneth Frank.)

far more prevalent on the urban beaches of a very populous coastline. Recreational fishing during grunion spawning runs can be extremely efficient and thorough, in spite of legal restrictions and protections. California grunion are especially vulnerable to catch because these fish come fully out of water on sandy beaches and can be taken by bare hands as they flop about on shore. The unique recreational fishery targets only the spawning runs. This species is very rarely caught or seen while at sea.

Regulations offer some protection to grunion *L. tenuis*, including a closed season and gear restrictions. Anglers are limited to the simple use of bare hands for capturing the fish out of water (Spratt, 1986). Adults require a fishing license, but children do not, and there are no bag limits (Gregory, 2001). The fishing regulations for this species were most recently revised in 1949, when the human population of California was a fraction of its current level, following a period of war when humans were scarce on beaches at night. Because the only time that this species is taken is during its actual spawning, the potential impact on the population is large (Sandrozinsky, 2013).

Enforcement of regulations on dark, deserted beaches occurs inconsistently. At times during open season the number of people fishing for California grunion is greater than the number of fish, and nearly every fish that appears is caught before any spawning can occur. The act of chasing *L. tenuis* to catch them, usually accompanied by shrieks, flashlights, and pounding feet racing along the beach, disrupts the run and may even stop it completely. The thrill is in the hunt, and although many grunion are caught, most anglers when queried do not have a plan for their catch. The fish simply suffocate in the buckets on shore, or die on the drive home. Surf fishermen collect *L. tenuis* to use later as bait.

The closed season for *L. tenuis* is supposed to provide a safe period for uninterrupted spawning in April and May (Spratt, 1986). However, in parts of the species range these months do not include the peak spawning runs. Changing temperatures or spring storms shift the spawning season so that the fixed dates of the closed season may not protect the largest runs. Recently the California grunion colonized disjunct bays north of Pt. Conception, including Monterey Bay, San Francisco Bay, and Tomales Bay, hundreds of kilometers north of their typical habitat in Southern California (Roberts et al., 2007; Johnson et al., 2009; Byrne et al., 2013). In each of these northern bays, the populations were very small, the spawning season was very short, and spawning starts so late the closed season of April was meaningless and May was just the beginning of the short spawning season (Martin et al., 2013). The closed season did not protect these fish from capture during their peak reproductive season in June and July, and only a little spawning occurred during April or May. Seasonal closures must be adjusted when habitats expand or shift during periods of climate change.

Another unique fishery on a beach-spawning fish is found in coastal New Zealand for the whitebait, *Galaxias maculatus*. This catadromous fish spawns in estuaries (McDowall, 1968; see Chapter 4 for details). After hatching, the larvae begin their migration upstream, where many are netted by fishermen to serve in a strange local cuisine. The larvae are stored in bottles in a little water, sold fresh by the liter as a slurry. The chef prepares the whitebait by pouring a glob of them into a bowl, mixing with raw eggs, then sautéing the resultant mixture like scrambled eggs. The result is then made into a sandwich with white bread and mayonnaise (McDowall, 1991). Beloved by Kiwis, this dish is quite possibly the very definition of an acquired taste (Figure 6.5).

FIGURE 6.5 (See color insert.) Recent hatchlings of the whitebait *Galaxias aculeatus* are the basis of this unique sandwich, served fresh seasonally in New Zealand. (Photo by Karen Martin.)

One reason that fisheries affect the future of beach-spawning fishes is that this is a way of indicating their economic value. However, there are other ways to assess the value of these fishes besides through direct take by humans. There are no fisheries on Pacific sandlance *Ammodytes hexapterus*, but they are considered ecologically valuable for their place in the food chain as fodder for salmon. Their cryptic habits surely have helped to protect them from fishing, with spawning having been first observed only in the past decade.

6.7   PARASITES ATTACK BEACH-SPAWNING FISHES FROM WITHIN

Animals in the wild often carry heavy burdens of parasites. Among beach-spawning fishes, the California grunion *L. tenuis* has had more than 25 parasites described (for example, see A. Olson, 1955, 1972, 1975, 1976, 1978; and L. Olson, 1979). According to Andrew Olson, Jr. (personal communication), no grunion is free of parasites. However, it is not clear if there is any effect of beach spawning, either positive or negative, on the parasite load of individual fishes or across species.

On a smaller scale, if eggs are damaged, bacteria may be able to penetrate an egg, followed by nematode worms *Rhabditis* sp. that may destroy the embryo of *L. tenuis* and fill up the egg with dozens of writhing worms (personal observations). Nematodes probably follow the scent of carbon dioxide emitted by the bacteria metabolism to enter the egg through the micropyle, and then, once inside, are unable to escape (Michael McClure, personal communication).

The inadvertent but sometimes devastating impacts of other anthropogenic activities on shore will be addressed in Chapter 9. See Chapter 10 for conservation efforts people are making to protect these vulnerable species.

So why do so many species of fishes have sex on the beach? Obviously, this does not avoid predation. The adults are challenged by a gauntlet of predators at sea when they aggregate before the run, and by marine and terrestrial predators during the run in shallow water and on land. Even the defenseless eggs fall prey to opportunistic birds and mammals and to slow-moving but determined worms and maggots during their incubation. The edge of the sea can be a treacherous place to begin life, although perhaps the predation is concentrated around the time of spawning, and less likely thereafter. Chapter 7 examines how embryos adapt to the physiological challenges of incubation on the beach.

REFERENCES

Able, K. W. & Hata, D. (1984). Reproductive behavior in the *Fundulus heteroclitus–F. grandis* complex. *Copeia* 1984, 820–825.

Allen, L. A. & Horn, M. H. (2006). *The Ecology of Marine Fishes: California and Adjacent Waters*. Berkeley: University of California Press, 670 pp.

Batts, B. S. (1961). Intertidal fishes as food of the common garter snake. *Copeia* 1961, 350–351.

Broughton, J. M., Cannon, V. I. & Arnold S. (2006). The taphonomy of owl-deposited fish remains and the origin of the Homestead Cave ichthyofauna. *Journal of Taphonomy* 4, 69–95.

Byrne, R. J., Bernardi, G. & Avise, J. (2013). Spatiotemporal genetic structure in a protected marine fish, the California grunion (*Leuresthes tenuis*), and relatedness in the genus *Leuresthes*. *Journal of Heredity* 104, 521–531.

Carscadden, J. E. (1981). *Underwater World: Capelin*. Communications Directorate. Ottawa, ON: Department of Fisheries and Oceans Canada.

Carscadden, J. E., Frank, K. T. & Miller, D. S. (1989). Capelin (*Mallotus villosus*) spawning on the southeast shoal: influence of physical factors past and present. *Canadian Journal of Fisheries and Aquatic Sciences* 46, 1743–1754.

Cavanagh, J. W., Martinez, K. M., Higgins, A., & Horn, M. H. (2014). Does the beach-spawning grunion eat its own eggs? Eighth graders use inquiry-based investigations to collect real data in a university laboratory. *The American Biology Teacher* 76, 178–182. http://www.jstor.org/stable/10.1525/abt.2014.76.3.5.

Coleman, R. (1992). Reproductive biology and female parental care in the cockscomb prickleback, *Anoplarchus purpurescens* (Pisces: Stichaeidae). *Environmental Biology of Fishes* 41, 177–186.

Coll, M., Palomera, I., Tudela, S. & Sarda, F. (2006). Trophic flows, ecosystem structure and fishing impacts in the South Catalan Sea, Northwestern Mediterranean. *Journal of Marine Systems* 59, 63–96.

Conover, D. O. & Kynard, B. E. (1984). Field and laboratory observations of spawning periodicity and behavior of a northern population of the Atlantic silverside, *Menidia menidia*. *Environmental Biology of Fishes* 11, 161–171.

Cucherousset, J., Bouletreau, S., Azemar, F., Compin, A., Gillaume, M. & Santoul, F. (2012). "Freshwater killer whales": beaching behavior of an alien fish to hunt land birds. *PLOS One* DOI: 10.1371/journal.pone.0050840.

DeMartini, E. E. (1976). *The Adaptive Significance of Territoriality and Egg Cannibalism in the Painted Greenling, Oxylebius Pictus Gill, a Northeastern Pacific Marine Fish*. PhD Thesis. University of Washington.

Duffy-Echevarria, E. E., Connor, R. C. & St. Aubin, D. J. (2008). Observations of strand-feeding behavior by bottlenose dolphins (*Tursiops truncatus*) in Bull Creek, South Carolina. *Marine Mammal Science* 24: 202–206. DOI: 10.1111/j.1748-7692.2007.00151.x.

Duran, J. (2008). Eulachon. In *Status of the Fisheries Report*, 7.1–7.4. Sacramento: California Department of Fish and Game.

Elliott, M. L., Hurt, R. & Sydeman, W. J. (2007). Breeding biology and status of the California least tern *Sterna antillarum browni* at Alameda Point, San Francisco Bay, California. *Waterbirds* 30, 317–325.

Feder, H. M., Turner, C. H. & Limbaugh, C. (1974). Observations on fishes associated with kelp beds in southern California. *Fish Bulletin* 160. State of California Resources Agency, California Department of Fish and Game, 144 pp.

FitzGerald, G. J. & van Havre, N. (1987). The adaptive significance of cannibalism in sticklebacks (Gasterosteidae: Pisces). *Behavioral Ecology and Sociobiology* 20, 125–128.

Gallup, F. N. (1949). Banding recoveries of *Tyto alba*. *Bird-Banding* 20, 150.

Gjøsæter, H. (1998). The population biology and exploitation of capelin (*Mallotus villosus*) in the Barents Sea. *Sarsia* 83, 453–496.

Gladstone, W. (1987). The eggs and larvae of the sharpnose pufferfish *Canthigaster valentini* (Pisces: Tetraodontidae) are unpalatable to other reef fishes. *Copeia* 1987, 227–230.

Goncalves, E. J. & Almada, V. C. (1998). A comparative study of territoriality in intertidal and subtidal blennioids (Teleostei, Blennioidei). *Environmental Biology of Fishes* 51, 257–264.

Gregory, P. (1978). Feeding habits and diet overlap of three species of garter snakes (*Thamnophis*) on Vancouver Island. *Canadian Journal of Zoology* 56, 1967–1974.

Gregory, P. A. (2001). Grunion. In *California's Living Marine Resources: A Status Report* (Leet, W. S., Dewees, C. M., Klingbeill, R. & Larson, E. J., eds.), 246–247. Sacramento, CA: California Department of Fish and Game.

Griem, J. N. & Martin, K. L. M. (1997). Predatory birds are present on the beach before and during California grunion runs. *American Zoologist* 37, 82A.

Haegle, C. W. (1993a). Seabird predation of Pacific herring, *Clupea pallasi*, spawn in British Columbia. *The Canadian Field-Naturalist* 107, 73–82.

Haegle, C. W. (1993b). Epibenthic invertebrate predation of Pacific herring, *Clupea pallasi*, spawn in British Columbia. *The Canadian Field-Naturalist* 107, 83–91.

Hardwick, J. E. (1973). Biomass estimates of spawning herring, *Clupea harengus pallasi*, herring eggs, and associated vegetation in Tomales Bay. *California Fish & Game* 59, 36–61.

Hourston, A. S. & Haegle, C. W. (1980). Herring on Canada's Pacific Coast. *Canada Special Publication in Fisheries and Aquatic Sciences* 48, 23 pp.

Johnson, P. B., Martin, K. L., Vandergon, T. L., Honeycutt, R. L., Burton, R. S. & Fry, A. (2009). Microsatellite and mitochondrial genetic comparisons between northern and southern populations of California grunion *Leuresthes tenuis*. *Copeia* 2009, 467–476.

Kaschner, K., Karpouzi, V., Watson, R. & Pauly, D. (2006). Forage fish consumption by marine mammals and seabirds. In *On the Multiple Uses of Forage Fish: From Ecosystems to Markets* (Alder, J. & Pauly, D., eds.), 33–46. Fisheries Centre Research Reports 14(3). British Columbia, Canada: Fisheries Centre, University of British Columbia.

Katzir, G. & Martin, G. R. (1998). Visual fields in the black-crowned night heron. *IBIS* 140, 157–162. DOI: 10.1111/j.1474-919X.1998.tb04554.x.

Lassuy, D. R. & Moran, D. (1989). Species profiles: life histories and environmental requirements of coastal fishes and invertebrates (Pacific Northwest)—Pacific herring. *U.S. Fish and Wildlife Service Biological Reports* 82 (11.126). U.S. Army Corps of Engineers, TR-EL-82-4.

Lillywhite, H. B., Sheehy III, C. M. & Zaidan III, F. (2008). Pitviper scavenging at the intertidal zone: an evolutionary scenario for invasion of the sea. *BioScience* 58, 947–955.

Lindstrom, K. & Ranta, E. (1992). Predation by birds affects population structure in breeding sand goby, *Pomatoschistus minutus*, males. *Oikos* 64, 527–532.

Lopez, J. C. & Lopez, D. (1985). Killer whales (*Orcinus orca*) of Patagonia, and their behavior of intentional stranding while hunting nearshore. *Journal of Mammalogy* 66, 181–183.

MacGinitie, G. E. & MacGinitie, N. (1949). *Natural History of Marine Animals*. New York: McGraw-Hill, 473 pp.

Manica, A. (2002). Filial cannibalism in telost fish. *Biological Reviews* 77, 262–277.

Martin, K. L. M. & Raim, J. G. (2014). Avian predators target nocturnal beach spawning runs of the marine fish, California grunion, *Leuresthes tenuis* (Atherinopsidae) at Malibu Lagoon. *Bulletin Southern California Academy of Sciences* 113 (in press).

Martin, K. L., Hieb, K. A. & Roberts, D. A. (2013). A Southern California icon surfs north: local ecotype of California grunion, *Leuresthes tenuis* (Atherinopsidae) revealed by multiple approaches during temporary habitat expansion into San Francisco Bay. *Copeia* 2013, 729–739. DOI: 10.1643/CI-13-036.

Martin, K. L., Martin, A. D., Martin, G. D. and Murrie, M. (2011). *Surf, Sand & Silversides: The California Grunion*. Short documentary. Malibu, CA: Malibu Moondance Production.

McDowall, R. M. (1968). *Galaxias maculatus* (Jenyns), the New Zealand whitebait. *New Zealand Marine Department of Fisheries Research Bulletin* 2, 1–84.

McDowall, R. M. (1991). *Conservation and Management of the Whitebait Fishery*. Science and Research Series No 38. New Zealand, Wellington: Department of Conservation.

McGowan, C. P., Hines, J. E., Nichols, J. D., Lyons, J. E., Smith, D. R., Kalasz, K. S., Niles, L. J., Dey, A. D., Clark, N. A., Atkinson, P. W., Minton, C. D. T. & Kendall, W. (2011). Demographic consequences of migratory stopover: linking red knot survival to horseshoe crab spawning abundance. *Ecosphere* 2: art 69. http://dx.doi.org/10.1890/ES11-00106.1.

Middaugh, D. P. (1981). Reproductive ecology and spawning periodicity of the Atlantic silverside, *Menidia menidia* (Pisces: Atherinidae). *Copeia* 1981, 766–776.

Middaugh, D. P., Kohl, H. W. & Burnett, L. E. (1983). Concurrent measurement of intertidal environmental variables and embryo survival for the California grunion, *Leuresthes tenuis*, and Atlantic silverside, *Menidia menidia* (Pisces: Atherinidae). *California Fish & Game* 69, 89–96.

Moyer, J. T. (1987). Quantitative observations of predation during spawning rushes of the Labrid fish *Thalassoma cupido* at Miyake-jima, Japan. *Japanese Journal of Ichthyology* 34, 76–81.

Olson, Jr., A. C. (1950). Ground squirrels and horned larks as predators upon grunion eggs. *California Fish & Game* 36, 323–327.

Olson, Jr., A. C. (1955). *Parasites of the Grunion*, Leuresthes tenuis (Ayres). PhD Dissertation. Oregon State College, 36 pp.

Olson, Jr., A. C. (1972). *Argulus melanostictus* and other parasitic crustaceans on the California grunion, *Leuresthes tenuis* (Osteichthyes: Atherinidae). *Journal of Parasitology* 58, 1201–1204.

Olson, Jr., A. C. (1975). Metacercaria of *Bucephalopsis labiatus* (Trematoda: Bucephalidae) in the California grunion, *Leuresthes tenuis*. *Journal of Parasitology* 61, 50.

Olson, Jr., A. C. (1976). *Asymphylodora atherinopsidis* (Trematoda: Monorchiidae) from the California grunion, *Leuresthes tenuis*, including a redescription. *Journal of Parasitology* 63, 295–298.

Olson, Jr., A. C. (1978). *Lepocreadium manteri* sp. N. (Trematoda: Lepocreadiidae) from the California grunion, *Leuresthes tenuis*, and its hyperparasitic microsporidan. *Proceedings Helminthological Society Washington* 45, 155–157.

Olson, L. J. (1979). *Host-Parasite Associations of the Grunions, Leuresthes sardina and Leuresthes tenuis from the Gulf of California*. PhD Dissertation. University of Arizona, 97 pp.

Pauly, D., Christensen, V., Dalsgaard, J., Froese, R. & Torres Jr., F. (1998). Fishing down marine food webs. *Science* 279, 860–863.

Perez, M. A. & Bigg, M. A. (1986). Diet of northern fur seals (*Callorhinus ursinus*) off western North America. *U.S. National Marine Fisheries Service Fisheries Bulletin* 84, 957–972.

Petersen, C. W., Mazzoldi, C., Zarrella, K. A. & Hale, R. E. (2005). Fertilization mode, sperm characteristics, mate choice and parental care patterns in *Artedius* spp. (Cottidae). *Journal of Fish Biology* 67, 239–254. DOI: 10.111/j.1095-8649.2005.00732.x.

Pfister, C. A. (2007). Sculpins. In *Encyclopedia of Tide Pools and Rocky Shores* (Denny, M. W. & Gaines, S., eds.), 485–486. Berkeley: University of California Press.

Piatt, J. F. & Anderson, P. (1996). Response of common murres to the *Exxon Valdez* oil spill and long-term changes in the Gulf of Alaska marine ecosystem. pp. 720–737 in: Rice, D., Spies, R. B., Wolfe, D. A. & and Wright, B. A. (eds.), *Proceedings of the Exxon Valdez Oil Spill Symposium.* American Fisheries Society Symposium 18.

Pillsbury, R. W. (1957). Avoidance of poisonous eggs of the marine fish *Scorpaenichthys marmoratus* by predators. *Copeia* 1957, 251–252.

Rich, C. & Longcore, T. (eds.). 2006. *Ecological Consequences of Artificial Night Lighting.* Washington, DC: Island Press, 458 pp.

Roberts, D. A., Lea, R. N. & Martin, K. M. (2007). First record of the occurrence of the California grunion, *Leuresthes tenuis*, in Tomales Bay, California; a northern extension of the species. *California Fish & Game* 93, 107–110.

Sancho, G. (2000). Predatory behaviors of *Caranx melampygus* (Carangidae) feeding on spawning reef fishes: a novel ambushing strategy. *Bulletin of Marine Science* 66, 487–496.

Sancho, G., Petersen, C. W. & Lobel, P. S. (2000). Predator-prey relations at a spawning aggregation site of coral reef fishes. *Marine Ecology Progress Series* 203, 275–288.

Sandrozinsky, A. (2013). *Status of the Fishery: California Grunion.* Status of the Fisheries Report. California Department of Fish and Wildlife. http://www.dfg.ca.gov/marine/status.

Santos, C. D., Miranda, A. C., Granadeiro, J. P., Lourenço, P. M., Saraiva, S. & Palmeirim, J. M. (2010). Effects of artificial illumination on the nocturnal foraging of waders. *Acta Oecologica* 36, 166–172. DOI:10.1016/j.actao.2009.11.008.

Smith, D. R., Jackson, N. L., Nordstrom, K. F. & Weber, R. G. (2011). Beach characteristics mitigate effects of onshore wind on horseshoe crab spawning: implications for matching with shorebird migration in Delaware Bay. *Animal Conservation* 14, 5, 575.

Spratt, J. D. (1986). *The amazing grunion.* Marine Resource Leaflet No. 3. California Department of Fish and Game, 7 pp.

Sweetnam, D. A., Baxter, R. D. & Moyle, P. B. (2001). True smelts. In *California's Living Marine Resources: A Status Report,* 470–478. Sacramento: California Department of Fish and Game.

Takita, T., Middaugh, D. P. & Dean, J. M. (1984). Predation of a spawning atherinid fish *Menidia menidia* by avian and aquatic predators. *Japanese Journal of Ecology* 34, 431–437.

Taylor, F. H. C. (1971). Variation in hatching success in Pacific herring (*Clupea pallasii*) eggs with water depth, temperature, salinity, and egg mass thickness. *Rapp. Proces-Verbaux Reunions Cons. Perm. Int. Explor. Mer* 160, 34–41.

Tewksbury, H. T. & Conover, D. O. (1987). Adaptive significance of intertidal egg deposition in the Atlantic silverside *Menidia menidia. Copeia* 1987, 76–83.

Thomson, D. A. & Muench, K. A. (1976). Influence of tides and waves on the spawning behavior of the Gulf of California grunion, *Leuresthes sardina* (Jenkins and Evermann). *Bulletin Southern California Academy of Science* 75, 198–203.

Vermeer, K. & Devito, K. (1986). Size, caloric content, and association of prey fishes in meals of nestling Rhinoceros Auklets. *Murrelet* 67, 1–9.

Walker, B. W. (1949). *The periodicity of spawning by the grunion, Leuresthes tenuis, and atherine fish.* Doctoral Dissertation. Los Angeles, CA: University of California, Scripps Institution of Oceanography. http://www.escholarship.org/uc/item/1j33928x.

Walker, B. W. (1952). A guide to the grunion. *California Fish & Game* 38, 409–420.

Yamahira, K. (1996). The role of intertidal egg deposition on survival of the puffer, *Takifugu niphobles* (Jordan et Snyder), embryos. *Journal of Experimental Marine Biology and Ecology* 198, 291–306.

7 Beach Babes: Terrestrial Incubation and Beach-Spawning Fishes

The earliest stages of life can be the most vulnerable. This is when vast changes occur, as a new individual proceeds through a program of development from a single cell to a multicellular organism with tissues and organs. Differentiation into muscles, sense organs, nervous tissue, and blood all takes place while the embryo is unable to move independently to escape a difficult environment or avoid predation. Beach-spawning fishes oviposit in the intertidal zone, a highly variable environment that is ruled by tidal cycles of inundation and air exposure. Placing one of the most vulnerable stages of life in this zone of constant change is a deliberate choice that deserves some careful examination. This chapter focuses on the effects of beach spawning on the early life of the many diverse species of fishes with these amphibious embryos.

7.1 DEMERSAL EGGS ARE WELL-SUITED FOR BEACH SPAWNING

Beach-spawning fishes have demersal eggs that are denser than water and sink to the substrate. In contrast, most species of marine fishes produce transparent floating pelagic eggs (Kendall et al., 1984; Pauly & Pullin, 1988). Lynne Parenti (personal communication) has identified several distinctions between pelagic and demersal eggs, in addition to whether they float or sink (Table 7.1).

Demersal eggs of marine fishes in general are larger than pelagic eggs and have a "stickiness" that allows attachment to vegetation, rocks, or other substrates, or to siblings' eggs within the clutch (Rizzo et al., 2002). During oviposition, demersal eggs may be attached to vegetation or rocks (Coleman, 1999) or adhere to other eggs to consolidate the clutch (Strathmann & Hess, 1999). Eggs may be hidden by nesting in cavities or crevices, by burying eggs in sand or gravel, or by internal gestation within the female's body or a male's brood pouch. Eggs that are not strongly adherent may be placed carefully into nests or simply settle into crevices of gravel or sand with water flow. Some eggs may be nudged into position by one of the parents after spawning.

Pelagic eggs use hydrolysis of yolk macromolecules to increase the number of particles, and thereby increase osmolarity, to draw water into the egg as a means of achieving neutral buoyancy (Craik & Harvey, 1986, 1987). In contrast, demersal eggs have limited hydrolysis of the yolk at the time of final oocyte maturation (McPherson et al., 1989). This means the demersal eggs not only sink, but they are osmotically more similar to adult marine fishes than are other marine eggs (Fyhn et al., 1999). Freshwater fishes, hyperosmotic to their environment, tend to produce demersal eggs, and these are placed carefully within a limited area rather than broadcast as pelagic eggs are (Fuiman, 2002).

TABLE 7.1

Comparison of Pelagic and Demersal Eggs

Demersal	Pelagic
Sink	Float
Limited hydrolysis	Maximum hydrolysis
Thick Zona Pellucida	Thin Zona Pellucida
Large	Small
May have filaments	No filaments
Sticky	Smooth

Source: As categorized by Lynne Parenti (personal communication).

Egg size for pelagic fishes is related to many environmental and genetic parameters including the species, the amount of food recently available to the mother, and her own body size (Pauly & Pullin, 1988). For demersal fishes, egg size may also be constrained by the effects of position and flow of water and oxygen to eggs within a clutch (Petersen & Hess, 2011). The maternal investment to make larger eggs is mainly in the form of proteins and fats as support for metabolism and growth. During the time within the egg, the embryos are completely dependent on this stored nutrition and energy, until hatching out as larvae that can swim and find their own food.

Larger eggs allow longer incubation times of days or weeks before hatching, even for pelagic eggs (Pauly & Pullin, 1988). Pelagic eggs tend to be small; 70 percent are 1.5 mm in diameter or less (Ahlstrom & Moser, 1980). Demersal eggs tend to be larger. The largest teleost eggs known are demersal eggs of an intertidal zoarcid, the South American *Austrolycus depressiceps*, at 9.8 mm in diameter (Matallanas et al., 1990). In some species of teleosts, the incubation time may be extended even after hatching competence to allow an environmental cue to trigger hatching when conditions are right for larval emergence (Moffatt & Thomson, 1978; Martin, 1999; Martin et al., 2011).

7.2   THE CHORION SURROUNDS AND PROTECTS THE EMBRYO

The chorion, a tough membrane that surrounds the anamniotic egg, is the barrier between the embryo and the environment. It allows the exchange of oxygen and carbon dioxide and the movement of water in both directions, but resists the entry or release of other substances. This provides protection from infection or toxic chemicals. Movements of water and solutes are controlled to some extent through aquaporin channels and ion channels (Machado & Podrabsky, 2007). After fertilization, the membrane of the chorion of fishes absorbs water from the surrounding environment and "hardens" by means of protein rearrangements (Yamagami, 1988).

In eggs of beach-spawning fishes such as the capelin *Mallotus villosus* or the California grunion *Leuresthes tenuis* that spend much of their incubation emerged from water, the hardened chorion feels sturdy and stiff, similar to a tough plastic coating around the egg. We have unsuccessfully attempted to preserve intact embryos by dropping eggs of *L. tenuis* in formalin, a preservative that is commonly used for fishes. However, after several hours in the solution, the embryos remained alive and apparently unaffected. To physically open the chorion to extract an embryo before it is ready to hatch requires enzymatic softening for several hours and the use of a sharp scalpel or craft knife (Moravek & Martin, 2011).

The chorion may be thicker than usual in beach-spawning fishes (Moravek & Martin, 2011; Mourabit et al., 2011). The chorion of the capelin *Mallotus villosus* is tough and resists damage from the gravel beach when shuffled about by waves (Templeman, 1948). The ultrastructure of the chorion of the beach-spawning blenny *Andamia tetradactyla* reveals seven layers covered by an adhesive disc of clasping filaments (Shimizu et al., 2011) (Figure 7.1). The four-eyed

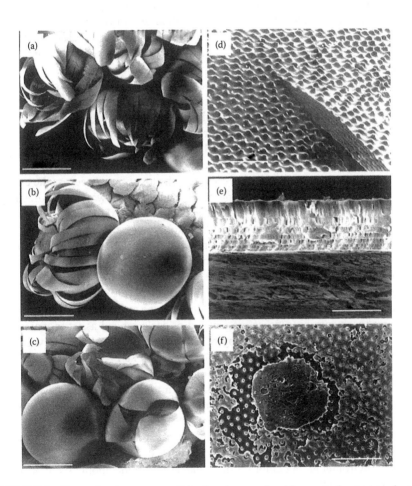

FIGURE 7.1   External egg structure of the beach-spawning blenny *Andamia tetradactyla* with SEM micrographs of broken membrane of eggs after hatching (a, b, c); broken edge of egg membrane after hatching (d); cross-section of egg membrane comprising seven continuous horizontal lamallae (e), and surface morphology of membrane (f). (From Shimizu, N. et al., *Japanese Journal of Ichthyology*, 58, 2011, 75–79. Used with permission of Norio Shimizu.)

blenny *Dialommus fuscus* is presumed to spawn intertidally, and its ova show thick filaments on the chorion (Howe, 1991). Intertidal gobiesocid clingfishes *Lepadogaster lepadogaster* and *Apletodon dentatus* have adhesive eggs with bifurcating filaments embedded in a mucous layer (Breining & Britz, 2000). A third sympatric clingfish, *Diplecogaster bimaculata*, has a similar chorion but does not live as high on shore. Of these three species, *L. lepadogaster* spawns highest in the intertidal zone and also has the largest eggs and the thickest multi-layered zona radiata.

Many species of amphibians incubate eggs terrestrially; however, these differ in several ways from teleost eggs (Martin & Carter, 2013). For one, amphibian eggs are substantially larger, and the species that produce terrestrial eggs have the largest of the amphibian eggs. For another, amphibian eggs are surrounded by a gel matrix that protects them from desiccation and ultraviolet exposure in air (Epel et al., 1999; Marco et al., 2001). Teleost eggs may have adherent filaments to be attached to rocks and substrates, but no gel or other matrix covers the surface of the chorion.

Most marine vertebrates, including teleosts, have lower osmolarity than sea-water; thus seawater is a desiccating environment. A tough chorion protects the embryo from water loss in seawater and also aids in resisting desiccation of the embryo when it is tidally exposed to air. Thus, this adaptation for marine life is also an adaptation that aids terrestrial life. During air exposure for the eggs of *Fundulus heteroclitus* the aquaporin channels of the chorion are down-regulated (Tingaud-Sequeira et al., 2009). This restricts the loss of water across the chorion. Embryos of *F. heteroclitus* incubating in air saturated with water vapor developed faster and were higher in water content and dry weight than a similar group of embryos that incubated in seawater (Tingaud-Sequeira et al., 2009). It is not known how well these eggs retain water in lower humidities; however, in nature the eggs may be sheltered by the habitat or tidally immersed in water so frequently that low humidity is not encountered.

Desiccated fish eggs appear dimpled and creased; severe desiccation can be fatal to intertidal eggs (Lee & Levings, 2007; Chuaypanang et al., 2013). Oviposition sites for many beach-spawning fishes are covered or enclosed and assist in maintaining humidity (DeMartini, 1988), either by burial under sand, attachment to vegetation, or hiding in crevices and empty shells (see Figure 3.1 and Figure 4.1).

The annual killifish *Austrofundulus limnaeus* is not a beach-spawning fish but it does incubate its eggs out of water, buried in mud left by a drying freshwater pool for months over a long dry season (Wourms, 1972). The eggs are able to maintain hydration, possibly by secreting substances into the perivitelline space surrounding the embryo that creates a barrier to prevent water loss to the sur-rounding mud (Podrabsky et al., 2001). This process also restricts any exchange with the environment, but embryos of *A. limnaeus* are metabolically depressed in a diapause during this period and require very little oxygen to survive (Podrabsky & Hand, 1999; Podrabsky et al., 2007). The perivitelline space in intertidal fish eggs is also fluid-filled. The possibility that it may function to adjust water loss or uptake for embryos of other species deserves further study.

In freshwater, the teleost egg tends to absorb water osmotically rather than losing it as in seawater. Swelling of the egg is prevented beyond a certain point by the capsule pressure of the chorion, analogous to the function of the cell wall in plant turgor. The in-between salinity of brackish water in estuaries may provide an osmotically hospitable zone for the nurseries of beach-spawning fishes, until the youngsters develop to the point of being able to osmoregulate like adults. Estuaries are nursery areas for larvae and juveniles of many species of fishes, and the origin of some beach-spawning species (Bamber & Henderson, 1988).

7.3   OVIPOSITION HEIGHT IS SPECIES-SPECIFIC AND MEDIATES BETWEEN OPPOSING RISKS OF AQUATIC HYPOXIA AND AERIAL DESICCATION

Beach-spawning fishes place their eggs in the intertidal zone at specific tidal heights that vary between species. From fertilization to hatching, site of oviposition is a major component of successful reproduction on beaches. Either too low or too high on shore can be fatal, but these optimal heights differ between species.

Oviposition height in the intertidal zone facilitates air exposure for the embryos of beach-spawning fishes. Duration and frequency of air exposure result from the ebb and flow of the tides and the topography of the nesting site. The variability of air temperatures and the potential for desiccation are far in excess of the variability of temperatures for ocean water during incubation, even in shallow tide pools of the intertidal zone. While risky, presumably the benefits of increased oxygen availability and the potential for increased temperatures outweigh the negative consequences for these species.

Access to oxygen is necessary for metabolism and development. Many beach-spawning fishes live in areas where aquatic hypoxia is likely to occur, such as tide pools and estuaries. To avoid aquatic hypoxia, estuarine fishes tend to spawn high in the intertidal zone, so the eggs are emerged frequently. Tide pool fishes may oviposit eggs in a monolayer so that each egg has maximal surface area exposed for diffusion. Using this strategy, eggs may be placed within a chamber, on the walls and roof of the chamber, so that as tides recede, the eggs become emerged into air. It is essential that mudskipper eggs avoid hypoxic water, or they will die (Etou et al., 2007).

Among resident intertidal fishes, most oviposit at the tidal height where they live (Martin et al., 2004). After hatching, larvae may enter the plankton. As they metamorphose from larvae to juveniles at settlement, young sculpins and blennies return to the intertidal zone; however, they settle into shallow pools higher in the intertidal zone than the adults, presumably to avoid predation (Zander et al., 1999).

Eggs of a few species do best when fully emerged throughout incubation (see Chapter 4). Other eggs and embryos survive at higher rates if emerged only occasionally. The vertical height of spawning in the intertidal zone is the major factor contributing to the frequency and duration of emergence for the eggs of fishes that migrate from offshore, but wave action, freshwater seeps, and biogenic habitat such as vegetation and empty shells can also alter the emergence status.

Oxygen availability and temperature are affected by duration and timing of air exposure, and both affect development rates and embryo survival (Nakashima & Wheeler, 2002; Smyder & Martin, 2002; Tingaud-Sequeira et al., 2009).

Fundulus heteroclitus (Fundulidae), the mummichog, spawns in the high intertidal zone of estuaries of the mid-Atlantic coast of North America. Its eggs are hidden in vegetation or within empty mussel shells in marshes. The female ovipositor on the anterior edge of the anal fin places eggs precisely (Taylor, 1999). Males lack this anatomical feature. See Chapter 3 for more on its spawning behavior. Eggs deeper in the water do not survive because of frequent exposure to aquatic hypoxia (Taylor & DiMichele, 1983).

Experimentally moving clutches of *F. heteroclitus* to different tidal heights significantly affected the survival of embryos (Taylor, 1999). None survived in the wild when submerged for the entire period of incubation. More recently a study in the laboratory showed eggs of *Fundulus heteroclitus* were able to survive during submergence in aerated water but developed at a slower rate than those eggs that incubated out of water (Tingaud-Sequeira et al., 2009). Out of water, the embryos rapidly transcribe a set of genes that appear to be similar to those for a response to desiccation stress, that may protect the embryo from reactive oxygen molecules during early development (Tingaud-Sequeira et al., 2013).

In the natural estuarine habitat, embryo survival for *F. heteroclitus* apparently depends on placing the eggs above the reach of hypoxic waters at low tides. The northern subspecies lays eggs in areas that are estuarine but less protected (Taylor, 1999). Their eggs are laid in algal mats or buried under sand high in the intertidal zone (Petersen et al., 2010).

Pacific herring *Clupea pallasii pallasii* (Clupeidae) eggs are attached to surfgrasses or macroalgae and may be exposed to air during low tides (Jones, 1972). If egg masses are large, they experience higher survival in air than in water, possibly because of the increased oxygenation and warmer temperatures that encourage development (Jones, 1972; Strathmann & Hess, 1999). Even when inundated by water, the eggs attached to the upper regions of these plants are closer to the surface and may have greater oxygen availability and thinner boundary layers than eggs attached on the same plant, but lower in the water column.

On days near the highest tides of full and new moons, kusafugu puffer *Takifugu niphobles* (Tetraodontidae) broadcasts nonadherent eggs during the lower of two daily high tides in summer (Uno, 1955). Eggs are exposed to air during incubation but must be inundated to hatch. Eggs almost never occur naturally in the low intertidal zone (Yamahira, 1996). However, when eggs were experimentally placed in lower zones of the beach, these embryos survived better than those naturally spawned in the highest intertidal zone. Survival of embryos during intertidal incubation varies over the spawning season (Yamahira, 1997). Higher in the intertidal zone, the typical placement by the spawning females, desiccation, and high temperatures cause loss of some incubating embryos (Yamahira, 1996). Thus it begs the question of why oviposition is so high.

The eggs of puffer fishes are toxic (Suyama & Uno, 1957), so the only predation risk for them at any tidal height is from congeners (see Chapter 6). The advantages of high intertidal spawning for this species may be the increased oxygen availability with greater times of emergence, coupled with the decreased likelihood of burial under sediments, and the avoidance of cannibalism (Yamahira, 1996).

Surf smelt *Hypomesus pretiosus* (Osmeridae) normally deposit their eggs high in the intertidal zone on gravel beaches, and these eggs are frequently emerged (Sweetnam et al., 2001). Surf smelt live in cool, temperate waters of Northern California and the Pacific Northwest. Spawning in this species may be so extensive that after the run, the beach literally is covered with its tiny, stalked eggs attached to gravel or rock. In this species, mortality is higher for eggs placed lower in the intertidal zone (Quinn et al., 2012). Surf smelt eggs suffer increased mortality from heat stress when shoreline modifications decrease the shade cover and alter the temperatures of the beach surface and interstitial spaces where the eggs reside (Rice, 2006). This highlights the special concerns for beach-spawning fishes during climate change; they will be affected not only by changes in water temperatures but also by increased air temperatures on shore during their critical embryonic stages.

Sympatric with the eggs of *H. pretiosus* but lower in the intertidal zone are the eggs of the Pacific sand lance *Ammodytes hexapterus* (Ammodytidae). The eggs of these two beach-spawning fishes form two distinct bands at different tidal heights on the shores of Puget Sound. This zonation emphasizes the species-specific nature of the preferred degree of exposure for fish embryos during incubation. These eggs are too small to be visible with the eye but can be obtained by careful washing of the gravel substrate with a strainer, followed by microscopic examination (Moulton & Penttila, 2001). Although tiny, the eggs are the target of a group of citizen scientists that collect samples to track spawning sites of these species for the resource managers in Washington State. See Chapter 10 for more on this effort.

The capelin *Mallotus villosus* (Osmeridae) spawns in the high intertidal zone during wind events that bring waves high on shore. Timing of spawning is not necessarily related to the syzygy high tides (Frank & Leggett, 1981a; Nakashima & Wheeler, 2002) and therefore not as predictable as in many other species. Small eggs are tightly attached individually to vegetation or gravel. These eggs remain stuck like glue while waves and sunlight play upon them during their incubation. Unlike most of the other species that spawn on beaches, embryos of *M. villosus* hatch in the gravel interstices, even if still separated from the ocean above the water line. Hatchlings may remain on the beach in the spaces between gravel for up to six days until freed by high waves generated by onshore winds (Frank & Leggett, 1981b). The larvae suffer high mortality after hatching while stranded on shore as they cannot swim or feed until returned to water.

Some populations of *M. villosus* spawn in deeper water (Carscadden et al., 1991). It appears that increased air temperatures would harm embryos on beaches in late summer, so the spawning capelin visit waters up to 60 m deep. In 2002, none of these subtidal embryos survived, while in comparison there was high survival for the intertidal embryos (Nakashima & Wheeler, 2002). There is no genetic divergence between capelin that spawn on beaches and those that spawn subtidally (Carscadden et al., 1991). Some subtidal sites are adjacent to beach-spawning sites, but others may be locations of former beaches that were inundated with sea level rise after the melting of the ice sheets of the Pleistocene. The effect of a beach-spawning site on embryo survival is important for considering future effects of climate change and sea level rise on other beach-spawning fishes (Nakashima & Taggart, 2002).

Demersal eggs of the Atlantic silverside *Menidia menidia* (Atherinopsidae) are attached by slender threads onto cordgrass, detrital mats, or abandoned crab burrows (Middaugh, 1981). Those embryos that are placed higher in the intertidal zone have fewer predators (Tewksbury & Conover, 1987) but may be troubled by decreased survival due to desiccation (Middaugh et al., 1983). These trade-offs may be site specific or stochastic, and if so, may not have much influence on parental choice of oviposition site.

Both California grunion *Leuresthes tenuis* (Atherinopsidae) and its congener, the Gulf grunion *L. sardina,* bury their nonadherent eggs in beach sand high in the intertidal zone while semilunar high tides are falling after a new or full moon (Walker, 1952; Thomson & Muench, 1976). Eggs are placed so that as tides fall, the embryos will incubate fully out of water (Martin et al., 2009). Embryos of either species will die if fully submerged for any length of time while buried. On the other hand, under natural conditions with burial in moist sand above the water line, survival of these grunion embryos approaches 97 percent (Middaugh et al., 1983), a remarkably high rate for wild fishes.

Eggs are emerged from water occasionally for short periods during development for many fishes that spawn in the high subtidal or low intertidal zone, whether or not one considers them to be beach-spawning fishes (Martin & Strathmann, 1999). These include some Scorpaeniformes that are not found in the intertidal zone, for example the sailfin sculpin *Nautichthys oculofasciatus* (Marliave, 1981) in the family Hemitripteridae, and two Cottidae, the red Irish lord *Hemilepidotus hemilepidotus* (Hart, 1973) and the cabezon *Scorpaenichthys marmoratus* (Pillsbury, 1957). The cabezon is often seen in kelp beds and caught off piers or from shore, indicating it usually lives subtidally. Cabezon eggs are poisonous (Hart, 1973), like the puffers, so placement higher on shore is not necessary to prevent egg depredation.

7.4   EMBRYOS ON BEACHES MAY INCUBATE FOR LONG PERIODS OF TIME

Pelagic eggs are relatively tiny and tend to hatch quickly after a short period of incubation, often in as little as one or two days (Pauly & Pullen, 1988). After such a short time, hatchlings are at an early stage of development, and at first these prolarvae may not even be able to feed or to swim. In contrast, embryos in demersal eggs may have longer incubation times of a week or more in either freshwater or seawater (Blaxter, 1988; Parenti, 1993; Martin et al., 2009).

In general, embryos from demersal eggs hatch at a more advanced larval stage. Some even bypass the larval stage completely, as in many elasmobranchs (Breder & Rosen, 1966). Because they are larger than pelagic eggs, demersal eggs have greater stored energy in the form of yolk proteins and lipids. This large yolk supply allows the longer incubation duration to a later stage of development and supports a larger size of hatchling before the need to feed (Matallanas et al., 1990; Martin et al., 2009).

For fishes with the ability to extend incubation beyond hatching competence, the larger egg size may also assist with prolonging the embryonic period (Martin & Carter, 2013).

Considering that pelagic eggs are tasty little morsels floating helplessly in the plankton, it makes sense for species with pelagic eggs to keep this vulnerable period of life as brief as possible. In contrast, the larger demersal eggs are often protected in some way during incubation. Some species hide the eggs during oviposition into crevices or under substrates. Many species with demersal eggs guard their nests, providing additional security for embryo survival. For example, the largest marine teleost eggs known, those of the zoarcid *Austrolycus depressiceps*, are found in rocky intertidal zone of Tierra Del Fuego, Argentina, and guarded by the female parent (Matallanas et al. 1990). The large eggs with high maternal investment in yolk in zoarcids contribute to very long incubation times, several months in some cases. Hatchlings are at an advanced stage of development (Matellanas et al., 1990; Ferry-Graham et al., 2007).

Large-yolked eggs, long incubation times, and hatchlings that bypass the larval stage are seen in some terrestrially breeding amphibians as well (Shine, 1978, 1989; Martin & Carter, 2013). Elasmobranchs also produce a few large eggs that produce juveniles rather than larvae upon hatching (Love, 2011). More comparative work examining the stage of development at the time of hatching competence is needed for beach-spawning fishes.

7.5   SOME BEACH-SPAWNING FISHES HAVE ENVIRONMENTAL SEX DETERMINATION

In many vertebrates, sex determination occurs at fertilization from the chromosomes donated by the parents. In others, gender is determined by some aspect of the environment during early life. Some species of teleosts have gender determined by the temperature of the incubation environment for the embryos or larvae, rather than by their DNA. Environmental sex determination may constrain habitat range by altering sex ratios in latitudinal extremes or at cooler depths.

Environmental sex determination is driven by temperature in the beach-spawning Atlantic silverside *Menidia menidia* (Conover & Kynard, 1981). However, genetic influences allow the production of at least a few of the opposite gender at any temperature. Conover et al. (1992) showed rapid adaptation in the laboratory over several generations of altered temperatures, to result in balanced sex ratios for *M. menidia* raised at a variety of constant temperatures. Variability in changing natural temperature regimes during the slower, seasonal reproduction that occurs in the wild could produce other effects that are not known. Estuarine populations of *M. peninsulae* from different areas vary in their response to temperature and sex determination (Yamahira & Conover, 2003). Sex determination in the California grunion (*Leuresthes tenuis*) may be influenced by temperature & photoperiod, as well as genetics (Brown et al., 2014).

Temperature-sensitive sex determination has not been studied in other species of beach-spawning fishes, although it may be present among the several species within the same family. Environmental sex determination has potential to create a detrimental impact of climate change on beach-spawning fishes in the near future (Conover et al., 1992). The effect of warming of sandy beaches is already a concern for sea turtles (LeBlanc et al., 2012).

7.6 EMBRYOS DEVELOP TERRESTRIALLY TO HATCHING COMPETENCE

Early development in fishes is not uniform. Differences exist between species for stage at hatching, rate of development, and order of appearance of some developmental landmarks among fish embryos of many types (White et al., 1984; Kimmel et al., 1995; Iwamatsu, 2004; Martin et al., 2009; Moravek & Martin, 2011). It is unfortunate that the unification of fish embryological staging is not as simple as the Gosner stages for the more uniformly developing frog tadpoles (Gosner, 1960). Thus, staging series for fishes tend to be species- or taxon-specific (Shardo, 1995). Five periods that always occur and follow the same order have been suggested by developmental ichthyologists (Shardo, 1995; Hill & Johnston, 1997; Fujimoto et al., 2006). These include (1) initiation of cleavage, (2) blastula formation, (3) gastrulation, (4) appearance of first somites, and (5) hatching competence. Useful as these are, they do not indicate the timing or order of organogenesis and fin bud formation, as these vary between species.

Embryonic staging series have been published for several beach-spawning fishes including the mummichog *Fundulus heteroclitus* (Armstrong & Child, 1965), the California grunion *Leuresthes tenuis* (David, 1939; Martin et al., 2009), and the self-fertilizing killifish *Kryptolebias marmoratus* (Mourabit et al., 2011), presumed to spawn on beaches of estuaries. In each case the terrestrial embryonic development proceeds in a manner similar to that of aquatic embryonic development, although faster in air (Tingaud-Sequeira et al., 2009, 2013), and of course more rapid with increasing temperatures (Smyder & Martin, 2002).

In addition to genetic programming, the developmental timetables of ectothermic vertebrates are affected by other environmental variables such as temperature, salinity, and oxygen availability. Salinity may change temporarily in the intertidal zone as a result of evaporation or rainfall, but the hardened chorion presumably prevents damage to the embryo during short-term exposure to altered salinities. However, consistent exposure to extremely high or low salinity causes deformities and mortality to embryos of the beach-spawning California grunion *Leuresthes tenuis* (Matsumoto & Martin, 2008). This type of consistent high salinity could result from brine disposal by desalination plants (see Chapter 9 for more on anthropogenic influences on beach-spawning grounds), and low salinity can occur for nests near freshwater outlets (Walker, 1952).

Surf smelt *Hypomesus pretiosus* spawn year-round in Puget Sound. In some areas in summer, the eggs overheat and there is almost 100 percent mortality. However in winter, even during a brief cold snap if the gravel freezes, the embryos are likely to survive with less than 5 percent mortality (Penttila, 2007).

Within species or between subspecies, a genetic component may influence developmental rate and growth in different ways (DiMichele & Powers, 1991; Williamson & DiMichele, 1997; Yamahira et al., 2007). For *Fundulus heteroclitus*, the subspecies from the more northern parts of the habitat range develop more rapidly at any temperature than fish from more southern areas at those same temperatures. Reaction norms allow latitudinal gradients so that all populations to hatch within 10–15 days, so the embryos are ready to hatch during the syzygy high tide that follows the semilunar tidal cycle when the spawning occurred (Conover, 1992; Yamahira et al., 2007). This type of latitudinal variation has been reported for the Atlantic silverside *Menidia menidia* and, to a lesser extent, for the Pacific coast *Atherinops affinis* (Baumann & Conover, 2011). On the other hand, the atherinopsid *Leuresthes tenuis*, a species that expanded its habitat range poleward just in the past decade (Roberts et al., 2007; Johnson et al., 2009; Martin et al., 2013), does not show latitudinal adaptation in development or growth at different temperatures (Brown et al., 2012).

Embryonic development is similar in form and duration for terrestrial incubation as compared with aquatic or marine incubation for fish eggs (Martin et al., 2009; Tingaud-Sequeira et al., 2009, 2013; Mourabit et al., 2011). See Table 7.2. With the large yolk and long duration of incubation, hatchlings are at a stage of development that in many cases is immediately ready to swim and feed in the plankton immediately, although hatchling *Porichthys notatus* remain in the nest for additional days (Crane, 1981), still attached by stalks to their boulder rest chamber.

One of the critical moments in an aquatic organism's life is the time of hatching from an immobile, encased embryo to a swimming, feeding larva (Sih & Moore, 1993). For beach-spawning fishes, this is the time that a terrestrially incubating embryo must become a marine larva, swimming in the ocean. An egg that hatches when it is out of water can be crushed or dried out, resulting in death of the embryo (Geiser & Seymour, 1989). Embryos of the capelin *Mallotus villosus* hatch as soon as they are competent developmentally to hatch, even if the larva is then stranded on shore (Frank & Leggett, 1981a). These terrestrial hatchlings shelter in the gravel interstices of the beach but lose condition rapidly because of starvation and must be released by waves within a few days or they will die.

Most teleost eggs avoid this fate because they must be submerged to hatch (Yamagami, 1988). Upon reaching hatching competence, if embryos of the kusafugu puffer *Takifugu niphobles* are out of water, they do not hatch (Yamahira, 1996). If tidal emergence is prolonged for too long around this time, the embryos may die without hatching; or, if they do hatch, the larvae die from desiccation. Embryonic California grunion *Leuresthes tenuis* also die without hatching if emerged, although they can prolong incubation for several weeks in the meantime (Martin, 1999; Smyder & Martin, 2002; Moravek & Martin, 2011).

7.7 SOME SPECIES WAIT FOR AN ENVIRONMENTAL CUE TO HATCH

The transition from enclosed embryo to swimming larva is especially challenging when this transition incorporates a change in habitat. Because the return of aquatic conditions is predictable from tidal ebb and flow, one way to avoid hatching at the wrong time is to rely on an environmental cue (Petranka et al., 1982; Martin et al., 2010; Warkentin, 2011; Martin et al., 2011). The variety of forms this alteration in developmental timetable takes in different species was recently reviewed for many different taxa (Warkentin, 2011).

TABLE 7.2

Comparison of Embryonic Development for Four Teleost Species

Species	L. tenuis	O. latipes	F. heteroclitus	D. rerio
Incubation temperature	20°C	26°C	20°C	26°C
Series	Atherinomorpha	Atherinomorpha	Atherinomorpha	Otophysi
Order	Atheriniformes	Beloniformes	Cyprinodontiformes	Cypriniformes
Family	Atherinopsidae	Adrianichthyidae	Fundulidae	Cyprinidae
Spawning habitat	Sandy beach, high intertidal	Freshwater	Estuary, high intertidal	Freshwater
Egg diameter (mm)	1.64 +/- 0.05	1.25 +/- 0.03	2	0.7
Per. 1: Initiation of cleavage	1 hpf	1.5 hpf	3 hpf	0.7 hpf
Per. 2: Blastula formation	8 hpf	6.5 hpf	11 hpf	2.2 hpf
Per. 3: Gastrulation	14 hpf	13 hpf	27 hpf	5.2 hpf
Neurulation	25 hpf	25 hpf	40 hpf	10 hpf
Per. 4: Appearance of first somites	32 hpf	27.5 hpf	52 hpf	10.3 hpf
Eye lenses	42 hpf	34 hpf	66 hpf	19.5 hpf
Heart beat initiation	45 hpf	44 hpf	74 hpf	24 hpf
Pectoral fin bud	58 hpf	58 hpf	84 hpf	30 hpf
Retinal pigment complete	95 hpf	106 hpf	192 hpf	30 hpf
Formation of swim bladder	106 hpf	101 hpf	at Period 5	39 hpf
Mouth opens and closes	154 hpf	192 hpf	at Period 5	after Period 5
Per. 5: Hatching competence	178 hpf	216 hpf	228 hpf	48–72 hpf
Potential extended incubation	to 34 d	NA	to 37 d	NA
Total embryonic period in days	7 (to 39) d	9 d	10 (to 37) d	2–3 d

HPF = hours post-fertilization

Data sources are as follows: for *L. tenuis*, Martin et al. (2009); for *O. latipes*, Iwamatsu (2004); for *F. heteroclitus*, Armstrong & Child (1965); for *D. rerio*, Kimmel et al. (1995).

Note: Two aquatic species, *Danio rerio* and *Oryzias latipes*, and two beach-spawning species with terrestrial development, *Leuresthes tenuis* and *Fundulus heteroclitus*, show similar progress and continuous development to hatching competence.

In some species of fishes with embryos that incubate terrestrially, hatching can be delayed after reaching developmental competence to hatch (Harrington & Haeger, 1958; Taylor et al., 1977; Darken et al., 1998; Martin, 1999; Martin & Carter, 2013). The embryo waits for a trigger, the environmental cue (Griem & Martin, 2000; Speer Blank & Martin, 2004). Until the environmental cue to hatch is present, the embryo remains within the egg. The metabolic consequences for survival and the maternal investment necessary for this strategy are discussed in Chapter 8.

REFERENCES

Ahlstrom, E. H. & Moser, H. G. (1980). Characters useful in identification of pelagic marine fish eggs. *California Cooperative Oceanic Fisheries Investigative Reports* 21, 121–131.

Armstrong, P. B. & Child, J. W. (1965). Stages in the normal development of *Fundulus heteroclitus*. *Biological Bulletin* 128, 143–168.

Bamber, T. N. & Henderson, P. A. (1988). Pre-adaptive plasticity in atherinids and the estuarine seat of teleost evolution. *Journal of Fish Biology* 33, 17–23.

Baumann, H. & Conover, D. O. (2011). Adaptation to climate change: contrasting patterns of thermal-reaction-norm evolution in Pacific versus Atlantic silversides. *Proceedings of the Royal Society B* 278, 2265–2273. DOI: 10.1098/rspb.2010.2479.

Blaxter, J. H. S. (1988). Pattern and variety in development. In *Fish Physiology* (Hoar, W. S. & Randall, D. J., eds.), Volume 11, Part A, 1–58. London: Academic Press.

Breder, C. M. & Rosen, D. E. (1966). *Modes of Reproduction in Fishes*. Garden City: Natural History Press, 941 pp.

Breining, T. & Britz, R. (2000). Egg surface structure of three clingfish species, using scanning electron microscopy. *Journal of Fish Biology* 56, 1129–1137. DOI: 10.1111/j.1095-8649.2000.tb02128.x.

Brown, E. E., Baumann, H. & Conover, D. O. (2012). Absence of countergradient and cogradient variation in an oceanic silverside, the California grunion *Leuresthes tenuis*. *Marine Ecology Progress Series* 461, 175–186. DOI:10.3354/meps09802.

Brown, E. E., Baumann, H. & Conover, D. O. (2014). Temperature and photoperiod effects on sex determination in a fish. *Journal of Experimental Marine Biology and Ecology* 461, 39–43.

Carscadden, J. E., Frank, K. T. & Miller, D. S. (1991). Capelin (*Mallotus villosus*) spawning on the southeast shoal: influence of physical factors past and present. *Canadian Journal of Fisheries and Aquatic Science* 46, 1743–1754.

Chuaypanang, S., Kidder, G. W. & Preston, R. L. (2013). Desiccation resistance in embryos of the killifish, *Fundulus heteroclitus*: single embryo measurements. *Journal of Experimental Zoology A: Ecological Genetics and Physiology* 319, 179–201.

Coleman, R. M. (1999). Parental care in intertidal fishes. In *Intertidal Fishes: Life in Two Worlds* (Horn, M. H., Martin, K. L. M. & Chotkowski, M. A., eds.), 165–180. San Diego: Academic Press.

Conover, D. O. (1992). Seasonality and the scheduling of life history at different latitudes. *Journal of Fish Biology* 41, 161–178. DOI: 10.1111/j.1095-8649.1992.tb03876.x.

Conover, D. O. & Kynard, B. E. (1981). Environmental sex determination: interaction of temperature and genotype in a fish. *Science* 213, 577–579.

Conover, D. O., Van Voorhees, D. A. & Ehtisham, A. (1992). Sex ratio selection and the evolution of environmental sex determination in laboratory populations of *Menidia menidia*. *Evolution* 46, 1722–1730.

Craik, J. C. A. & Harvey, S. M. (1986). Phosphorus metabolism and water uptake during final maturation of ovaries of teleosts with pelagic and demersal eggs. *Marine Biology* 90, 285–289.

Craik, J. C. A. & Harvey, S. M. (1987). The causes of buoyancy in eggs of marine teleosts. *Journal of the Marine Biological Association, United Kingdom* 67, 169–182.

Crane, J. M., Jr. (1981). Feeding and growth by the sessile larvae of the teleost *Porichthys notatus*. *Copeia* 1981, 895–897.

Darken, R. S., Martin, K. L. M. & Fisher, M. E. (1998). Metabolism during delayed hatching in terrestrial eggs of a marine fish, the grunion *Leuresthes tenuis*. *Physiological Zoology* 71, 400–406.

David, L. R. (1939). Embryonic and early larval stages of the grunion, *Leuresthes tenuis*, and of the sculpin, *Scorpaena guttata*. *Copeia* 1939, 75–80.

DeMartini, E. E. (1988). Spawning success in the male plainfin midshipman. I. Influences of male body size and area of spawning site. *Journal of Experimental Marine Biology and Ecology* 121, 177–192.

DiMichele, L. & Powers, D. A. (1991). Allozyme variation, developmental rate, and differential mortality in the teleost *Fundulus heteroclitus*. *Physiological Zoology* 64, 1426–1443.

Epel, D., Hemela, K., Shick, M. & Patton, C. (1999). Development in the floating world: defenses of eggs and embryos against damage from UV radiation. *American Zoologist* 39, 271–278.

Etou, A., Takeda, T., Yoshida, Y. & Ishimatsu, A. (2007). Oxygen consumption during embryonic development of the mudskipper (*Periophthalmus modestus*): implication for the aerial development in burrows. In *Fish Respiration and Environment* (Fernandes, M. N., Rantin, F. T., Glass, M. L. & Kapoor, B. G., eds.), 83–91. Enfield, NJ: Science Publishers.

Ferry-Graham, L. A., Drazen, J. C. & Franklin, V. (2007). Laboratory observations of reproduction in the deep-water Zoarcids *Lycodes cortezianus* and *Lycodapus mandibularis* (Teleostei: Zoarcidae). *Pacific Science* 61, 129–139. DOI: http://dx.doi.org/10.1353/psc.2007.0004.

Frank, K. T. & Leggett, W. C. (1981a). Wind regulation of emergence times and early larval survival in capelin (*Mallotus villosus*). *Canadian Journal of Fisheries and Aquatic Science* 38, 215–223.

Frank, K. T. & Leggett, W. C. (1981b). Prediction of egg development and mortality rates in capelin (*Mallotus villosus*) from meteorological, hydrographic, and biological factors. *Canadian Journal of Fisheries and Aquatic Science* 38, 1327–1338.

Fuiman, L. A. (2002). Special considerations of fish eggs and larvae. In *Fishery Science: The Unique Contributions of Early Life Stages* (Fuiman, L. E. & Werner, R. G., eds.), 1–32. Oxford, UK: Blackwell Science.

Fujimoto, T., Kataoka, T., Sakao, S., Saito, T., Yamaha, E. & Arai, K. (2006). Developmental stages and germ cell lineage of the loach (*Misgurnus anguillicaudatus*). *Zoological Science* 23, 977–989.

Fyhn, H. J., Finn, R. N., Reith, M. & Norberg, B. (1999). Yolk protein hydrolysis and oocyte free amino acids as key features in the adaptive evolution of teleost fishes to seawater. *Sarsia* 84, 451–456.

Geiser, F. & Seymour, R. S. (1989). Influence of temperature and water potential on survival of hatched, terrestrial larvae of the frog *Pseudophryne bibronii*. *Copeia* 1989, 207–209.

Gosner, K. L. (1960). A simplified table for staging anuran embryos and larvae with notes on identification. *Herpetologica* 16, 183–190.

Griem, J. N. & Martin, K. L. M. (2000). Wave action: the environmental trigger for hatching in the California grunion, *Leuresthes tenuis* (Teleostei: Atherinopsidae). *Marine Biology* 137, 177–181.

Harrington, R. W., Jr. & Haeger, J. S. (1958). Prolonged natural deferment of hatching in killifish. *Science* 128, 1511.

Hart, J. L. (1973). Pacific fishes of Canada. *Fisheries Research Board of Canada, Bulletin* 180, 1–740.

Hill, J. & Johnston, I. A. (1997). Photomicrographic atlas of Atlantic herring embryonic development. *Journal of Fish Biology* 51, 960–977. DOI: 10.1111/j.1095-8649.1997. tb01535x.

Howe, J. C. (1991). Egg surface morphology of *Dialommus fuscus* Gilbert (Pisces: Labrisomidae). *Journal of Fish Biology* 38, 149–152.

Iwamatsu, T. (2004). Stages of normal development in the medaka *Oryzias latipes. Mechanisms of Development* 121, 605–618.

Johnson, P. B., Martin, K. L., Vandergon, T. L., Honeycutt, R. L., Burton, R. S. & Fry, A. (2009). Microsatellite and mitochondrial genetic comparisons between northern and southern populations of California grunion *Leuresthes tenuis. Copeia* 2009, 467–476.

Jones, B. C. (1972). Effect of intertidal exposure on survival and embryonic development of Pacific herring spawn. *Journal of Fisheries Research Board, Canada* 29, 1119–1124.

Kendall, A. W., Ahlstrom, E. H. & Moser, H. G. (1984). Early life history stages of fishes and their characters. In *Ontogeny and Systematics of Fishes* (Moser, H. G., ed.), 11–22. Special Publication No. 1. Lawrence, KS: Allen Press, American Society of Ichthyologists and Herpetologists.

Kimmel, C. B., Ballard, W. W., Kimmel, S. R., Ullmann, B. & Schilling, T. F. (1995). Stages of embryonic development of the zebrafish. *Developmental Dynamics* 203, 253–310.

LeBlanc, A. M., Wibbels, T., Shaver, D. & Walker, J. S. (2012).Temperature-dependent sex determination in the Kemp's ridley sea turtle: effects of incubation temperatures on sex ratios. *Endangered Species Research* 19, 123–128. DOI: 10.3354/esr00465.

Lee, C. G. & Levings, C. D. (2007). The effects of temperature and desiccation on surf smelt (*Hypomesus pretiosus*) embryo development and hatching success: preliminary field and laboratory observations. *Northwest Science* 81, 166–171.

Love, M. S. (2011). *Certainly More Than You Want to Know about the Fishes of the Pacific Coast: A Postmodern Experience.* Santa Barbara, CA: Really Big Press, 672 pp.

Machado, B. E. & Podrabsky, J. E. (2007). Salinity tolerance in diapausing embryos of the annual killifish *Austrofundulus limnaeus* is supported by exceptionally low water and ion permeability. *Journal of Comparative Physiology (B)* 177, 809–820.

Marco, A., Lizana, M., Alvarez, A. & Blaustein, A. R. (2001). Egg-wrapping behaviour protects newt embryos from UV radiation. *Animal Behaviour* 61, 1–6.

Marliave, J. B. (1981). High intertidal spawning under rockweed, *Fucus distichus*, by the sharpnose sculpin, *Clinocottus acuticeps. Canadian Journal of Zoology* 59, 1122–1125.

Martin, K. L. M. (1999). Ready and waiting: delayed hatching and extended incubation of anamniotic vertebrate terrestrial eggs. *American Zoologist* 39, 279–288.

Martin, K. L. & Carter, A. L. (2013). Brave new propagules: terrestrial embryos in anamniotic eggs. *Integrative and Comparative Biology* 53, 233–247. DOI:10.1093/icb/ict018.

Martin, K. L. M. & Strathmann, R. (1999). Aquatic organisms, terrestrial eggs: early development at the water's edge. Introduction to the symposium. *American Zoologist* 39, 215–217.

Martin, K. L. M., Bailey, K., Moravek, C. & Carlson, K. (2011). Taking the plunge: California grunion embryos emerge rapidly with environmentally cued hatching. *Integrative and Comparative Biology* 51, 26–37. DOI: 10.1093/icb/icr037.

Martin, K. L. M., Heib, K. A. & Roberts, D. A. (2013). A Southern California icon surfs north: local ecotype of California grunion, *Leuresthes tenuis* (Atherinopsidae) revealed by multiple approaches during temporary habitat expansion into San Francisco Bay. *Copeia* 2013, 729–739. DOI: 10.1643/CI-13-036.

Martin, K. L., Moravek, C. L. & Flannery, J. A. (2009). Embryonic staging series for the beach spawning, terrestrially incubating California grunion *Leuresthes tenuis* with comparisons to other Atherinomorpha. *Journal of Fish Biology* 75, 17–38.

Martin, K. L. M., Moravek, C. L. & Walker, A. J. (2010). Waiting for a sign: extended incubation postpones larval stage in the beach spawning California grunion *Leuresthes tenuis* (Ayres). *Environmental Biology of Fishes* 91, 63–70. DOI: 10.1007/s10641-010-9760-4.

Martin, K. L. M., Van Winkle, R. C., Drais, J. E. & Lakisic, H. (2004). Beach-spawning fishes, terrestrial eggs, and air breathing. *Physiological and Biochemical Zoology* 77, 750–759.

Matallanas, J., Rucabado, J., Lloris, D. & Pilar Olivar, M. (1990). Early stages of development and reproductive biology of South American eelpout *Austrolycus depressiceps* Regan, 1913 (Teleostei: Zoarcidae). *Sciencias Marinas* 54, 257–261.

Matsumoto, J. K. & Martin, K. L. M. (2008). Lethal and sublethal effects of altered sand salinity on embryos of beach-spawning California grunion. *Copeia* 2008, 483–490. DOI: 10.1643/CP-07-097.

McPherson, R., Greeley, M. S. Jr., & Wallace, R. A. (1989). The influence of yolk protein proteolysis on hydration in the oocytes of *Fundulus heteroclitus*. *Development, Growth & Differentiation* 31, 475–483.

Middaugh, D. P. (1981). Reproductive ecology and spawning periodicity of the Atlantic silverside *Menidia menidia* (Pisces: Atherinidae). *Copeia* 1981, 766–776.

Middaugh, D. P., Kohl, H. W. & Burnett, L. E. (1983). Concurrent measurement of intertidal environmental variables and embryo survival for the California grunion, *Leuresthes tenuis*, and Atlantic silverside, *Menidia menidia* (Pisces: Atherinidae). *Califorina Fish & Game* 69, 89–96.

Moffatt, N. M. & Thomson, D. A. (1978). Tidal influence on the evolution of egg size in the grunions (*Leuresthes*, Atherinidae). *Environmental Biology of Fishes* 3, 267–273.

Moravek, C. L. & Martin, K. L. (2011). Life goes on: delayed hatching, extended incubation, and heterokairy in development of embryonic California grunion, *Leuresthes tenuis*. *Copeia* 2011, 308–314. DOI: 10.1643/CG-10-164.

Moulton, L. L. & Penttila, D. E. (2001). Field manual for sampling forage fish spawn in intertidal shore regions: *San Juan County Forage Fish Assessment Project*, Washington Department of Fish and Wildlife, Olympia, WA.

Mourabit, S., Edenbrow, M., Croft, D. P. & Kudoh, T. (2011). Embryonic development of the self-fertilizing mangrove killifish *Kryptolebias marmoratus*. *Developmental Dynamics* 240, 1694–1704.

Nakashima, B. S. & Taggart, C. T. (2002). Is beach-spawning success for capelin, *Mallotus villosus* (Muller), a function of the beach? *ICES Journal of Marine Science* 59, 897–908.

Nakashima, B. S. & Wheeler, J. P. (2002). Capelin (*Mallotus villosus*) spawning behavior in Newfoundland waters: the interaction between beach and demersal spawning. *ICES Journal of Marine Science* 59, 909–916.

Parenti, L. R. (1993). Relationships of atherinomorph fishes (Teleostei). *Bulletin of Marine Science* 52, 170–196.

Pauly, D. P. & Pullin, R. S. V. (1988). Hatching time in spherical, pelagic marine fish eggs in response to temperature and egg size. *Environmental Biology of Fishes* 22, 261–271.

Penttila, D. E. (2007). Marine forage fishes in Puget Sound. Technical Report 2007-03, Washington Department of Fish and Wildlife, Washington Sea Grant, Puget Sound Nearshore Partnership, Olympia, WA, 23 pp.

Petersen, C. W. & Hess, H. C. (2011). Evolution of parental behavior, egg size, and egg mass structure in sculpins. In *Adaptation and Evolution in Cottoid Fishes* (Goto, A., Munehara, H. & Yabe, M., eds.), 194–203. Kanagawa, Japan: Tokai University Press.

Petersen, C. W., Salinas, S., Preston, R. L. & Kidder, G. W. III. (2010). Spawning periodicity and reproductive behavior of *Fundulus heteroclitus* in a New England salt marsh. *Copeia* 2010, 203–210. DOI: http://dx.doi.org/10.1643/CP-08-229.

Petranka, J. W., Just, J. J. & Crawford, E. C. (1982). Hatching of amphibian eggs: the physiological trigger. *Science* 217, 257–259.

Pillsbury, R. W. (1957). Avoidance of poisonous eggs of the marine fish *Scorpaenichthys marmoratus* by predators. *Copeia* 1957, 251–252.

Podrabsky, J. E. & Hand, S. C. (1999). The bioenergetics of embryonic diapause in an annual killifish, *Austrofundulus limnaeus*. *Journal of Experimental Biology* 202, 2567–2580.

Podrabsky, J. E., Carpenter, J. F. & Hand, S. C. (2001). Survival of water stress in annual fish embryos: dehydration avoidance and egg envelope amyloid fibers. *American Journal of Physiology: Regulatory and Integrative Comparative Physiology* 280, R123–R131.

Podrabsky, J. E., Lopez, J. P., Fan, T. W., Higashi, R. & Somero, G. N. (2007). Extreme anoxia tolerance in embryos of the annual killifish *Austrofundulus limnaeus*: insights from a metabolomic analysis. *Journal of Experimental Biology* 210, 2253–2266.

Quinn, T., Krueger, K., Pierce, K., Penttila, D., Perry, K., Hicks, T. & Lowry, D. (2012). Patterns of surf smelt, *Hypomesus pretiosus*, intertidal spawning habitat use in Puget Sound, Washington State. *Estuaries and Coasts* 35, 1214–1228. DOI: 10.1007/s12237-012-9511-1.

Rice, C. A. (2006). Effects of shoreline modification on a northern Puget Sound beach: microclimate and embryo mortality in surf smelt (*Hypomesus pretiosus*). *Estuaries and Coasts* 29, 63–71.

Rizzo, E., Sato, Y., Barreto, B. P. & Godinho, H. P. (2002). Adhesiveness and surface patterns of eggs in neotropical freshwater teleosts. *Journal of Fish Biology* 61, 615–632. DOI: 10.1111/j.1095=6849.2002.tb00900.x.

Roberts, D., Lea, R. N. & Martin, K. L. M. (2007). First record of the occurrence of the California grunion, *Leuresthes tenuis*, in Tomales Bay, California; a northern extension of the species. *California Fish and Game* 93, 107–110.

Shardo, J. D. (1995). Comparative embryology of teleostean fishes. I. Development and staging of the American shad, *Alosa sapidissima* (Wilson, 1811). *Journal of Morphology* 225, 125–167.

Shimizu, N., Hara, M., Sakai, Y., Hashimoto, H. & Gushima, K. (2011). Ultrastructure of the surface morphology of eggs in the terrestrial spawning blenny, *Andamia tetradactyla* (Blenniidae). *Japanese Journal of Ichthyology* 58, 75–79.

Shine, R. (1978). Propagule size and parental care: the "safe harbor" hypothesis. *Journal of Theoretical Biology* 75, 417–424.

Shine, R. (1989). Alternative models for the evolution of offspring size. *American Naturalist* 134, 311–317.

Sih, A. & Moore, R. D. (1993). Delayed hatching of salamander eggs in response to enhanced larval predation risk. *American Naturalist* 143, 947–960.

Smyder, E. A. & Martin, K. L. M. (2002). Temperature effects on egg survival and hatching during the extended incubation period of California grunion, *Leuresthes tenuis*. *Copeia* 2002, 313–320.

Speer Blank, T. M. & Martin, K. L. M. (2004). Hatching events in the California grunion, *Leuresthes tenuis*. *Copeia* 2004, 21–27.

Strathmann, R. R. & Hess, H. C. (1999). Two designs of marine egg masses and their divergent consequences for oxygen supply and desiccation in air. *American Zoologist* 39, 253–260.

Suyama, M. & Uno, Y. (1957). Puffer toxin during the embryonic development of puffer, *Fugu (Fugu) niphobles* (Jordan et Snyder). *Bulletin of Japanese Society for Science and Fisheries* 23, 438–441.

Sweetnam, D. A., Baxter, R. D. & Moyle, P. B. (2001). True smelts. In *California's Living Marine Resources: A Status Report,* 470–478. Sacramento: California Department of Fish and Game.

Taylor, M. H. (1999). A suite of adaptations for intertidal spawning. *American Zoologist* 39, 313–320.

Taylor, M. H. & DiMichele, L. (1983). Spawning site utilization in a Delaware population of *Fundulus heteroclitus* (Pisces: Cyprinodontidae). *Copeia* 1983, 719–725.

Taylor, M. H., DiMichele, L. & Leach, G. J. (1977). Egg stranding in the life cycle of the mummichog, *Fundulus heteroclitus. Copeia* 1977, 397–399.

Templeman, W. (1948). The life history of the capelin (*Mallotus villosus*) in Newfoundland waters. *Bulletin of the Newfoundland Government Laboratory at St. John's* 17, 1–155.

Tewksbury, H. T. & Conover, D. O. (1987). Adaptive significance of intertidal egg deposition in the Atlantic silverside *Menidia menidia. Copeia* 1987, 76–83.

Thomson, D. A. & Muench, K. A. (1976). Influence of tides and waves on the spawning behavior of the Gulf of California grunion, *Leuresthes sardina* (Jenkins and Evermann). *Bulletin of the Southern California Academy of Science* 75, 198–203.

Tingaud-Sequeira, A., Zapater, C., Chauvigne, F., Otero, D. & Cerda, J. (2009). Adaptive plasticity of killifish (*Fundulus heteroclitus*) embryos: dehydration-stimulated development and differential aquaporin-3 expression. *American Journal of Regulatory and Integrative Comparative Physiology* 296, R1041–R1052. DOI: 10.1152/ajpregu91002.2008.

Tingaud-Sequeira, A., Lozano, J.-J., Zapater, C., Otero, D., Kube, M., Reinhardt, R. & Cerda, J. (2013). A rapid transcriptome response is associated with desiccation resistance in aerially-exposed killifish embryos. *PLOS One*, 8, 5, e64410.

Uno, Y. (1955). Spawning habit and early development of a puffer, *Fugu (Torafugu) niphobles* (Jordan et Snyder). *Journal of Tokyo University Fisheries* 41, 169–183.

Walker, B. (1952). A guide to the grunion. *California Fish & Game* 38, 409–420.

Warkentin, K. M. (2011). Environmentally cued hatching across taxa: embryos respond to risk and opportunity. *Integrative and Comparative Biology* 51, 14–25. DOI: 10.1093/icb/icr017.

White, B. N., Lavenberg, R. J. & McGowan, G. E. (1984). Atheriniformes: development and relationships. In *Ontogeny and Systematics of Fishes* (Moser, H. G., Richards, W. J., Cohen, D. M., Fahay, M. P., Kendall, Jr., A. W. & Richardson, S. L., eds.), 355–362. Special Publication 1. American Society of Ichthyologists and Herpetologists.

Williamson, E. G. & DiMichele, L. (1997). An ecological simulation reveals balancing selection acting on development rate in the teleost *Fundulus heteroclitus. Marine Biology* 128, 9–15.

Wourms, J. P. (1972). The developmental biology of annual fishes. III. Pre-embryonic and embryonic diapause of variable duration in the eggs of annual fishes. *Journal of Experimental Zoology* 182, 389–414.

Yamagami, K. (1988). Mechanisms of hatching in fish. In *Fish Physiology, Volume 11A* (Hoar, W. S. & Randall, D. J., eds.), 447–499. San Diego, CA: Academic Press.

Yamahira, K. (1996). The role of intertidal egg deposition on survival of the puffer, *Takifugu niphobles* (Jordan et Snyder), embryos. *Journal of Experimental Marine Biology and Ecology* 198, 291–306.

Yamahira, K. (1997). Hatching success affects the timing of spawning by the intertidally spawning puffer *Takifugu niphobles. Marine Ecology Progress Series* 155, 239–248.

Yamahira, K. & Conover, D. O. (2003). Interpopulation variability in temperature-dependent sex determination of the tidewater silverside *Menidia peninsulae* (Pisces: Atherinidae). *Copeia* 2003, 155–159.

Yamahira, K., Kawajiri, M., Takeshi, K. & Irie, T. (2007). Inter- and intrapopulation variation in thermal reaction norms for growth rate: evolution of latitudinal compensation in ectotherms with a genetic consraint. *Evolution* 61, 1577–1589.

Zander, C. D., Nieder, J. & Martin, K. L. M. (1999). Vertical distribution patterns. In *Intertidal Fishes: Life in Two Worlds* (Horn, M. H., Martin, K. L. M. & Chotkowski, M. A., eds.), 26–53. San Diego, CA: Academic Press.

8 Perilous Return to the Sea: Hatching after Beach Incubation

Hatching is a key life-history switch point (Sih & Moore, 1993). At hatching, an animal leaves the safety of its sessile, self-contained egg for the active freedom of larval life. For beach-spawning fishes, this transition must occur at a time when aquatic conditions prevail, as the delicate larvae are far less able to handle terrestrial conditions than the embryos wrapped in their sturdy chorions are.

A safe and successful transition back into a fully aquatic habitat is vital. For aquatic fish incubating on land, hatching at the wrong moment can be fatal. Environmental cues for hatching are important in many species and take different forms. However, there is at least one example of a beach-spawning species that hatches on a developmental timetable and does not require an environmental trigger.

Some fish embryos can prolong incubation by delaying hatching until conditions are favorable for the larvae. When the environmental cue to hatch is not present, incubation can be extended far beyond the time necessary to reach hatching competence. This chapter addresses how this alteration in developmental timetable helps or hinders the individual, the variety of forms this heterokairy takes in different species, the metabolic consequences, and the types of maternal investment necessary for this strategy.

8.1 INCUBATION ON BEACHES EXPOSES FISH EMBRYOS TO THE DANGER OF HATCHING ON LAND

The tough chorion that surrounds the egg provides protection to the embryo from desiccation and crushing. Eggs of fishes that spawn on gravel beaches and in sand can be tossed about by waves or walked upon by humans and other animals, usually without significant damage. Eggs of surf smelt can even endure brief bouts of freezing conditions. However, the situation changes once the embryo hatches out. The delicate hatchling requires aquatic conditions for buoyancy and to breathe. The larvae may need to feed soon after hatching, and that means small planktonic organisms must float nearby. Without its chorion and the cushion of fluid within the egg, the tiny hatchling can be injured or crushed by coarse sediments or suffocated by burial. Although exposure to high temperatures on a beach can kill embryos within eggs (Yamahira, 1997), overheating and desiccation risks are far greater for terrestrial larvae than for the egg-encased embryos.

However, embryos cannot remain in their eggs forever. The yolk is a limited source of energy that is consumed (Heming & Buddington, 1988). The size of the embryo may exceed the diffusion capacity of the chorion for oxygen delivery. Wastes or metabolic byproducts may accumulate. Most importantly, the genetic program for development is ready for the next stage of life. For most aquatic fishes, there is a range of times for development at a given temperature, but most hatching takes place within a fairly predictable window of time. For fishes incubating on land, the situation is not so simple.

Most fish eggs must be submerged in water to hatch (see below). This is not the case for the beach-spawning capelin *Mallotus villosus* (Osmeridae), however. Neither the spawning runs nor the hatching are synchronized by tides in this species (Frank & Leggett, 1981a,b; Nakashima & Wheeler, 2002). Embryos hatch on gravel beaches when the developmental time is right, regardless of the terrestrial conditions. Hidden in interstitial crevices to keep damp, the hatchlings are unable to feed and gradually lose condition until freed by wind waves that reach high enough on shore to release them into the sea. This species also spawns subtidally, and success in reproduction varies across beaches (Nakashima & Taggart, 2002).

Many amphibian species that incubate terrestrially are able to bypass the larval stage completely and hatch as a tiny version of the adult (Altig & McDiarmid, 2010). This feat avoids the danger of terrestrial life for larvae. One Australian toad *Pseudophryne bibroni*, with eggs that incubate terrestrially, still requires an aquatic tadpole stage. It is able to delay hatching until spring rains refill the pond, but sometimes eggs hatch before that and the larvae, like those of the capelin, survive briefly as terrestrial refugees but are not able to last long without the return of aquatic conditions in the form of rainfall (Geiser & Seymour, 1989).

8.2 HATCHLINGS MUST NAVIGATE THE PERILOUS RETURN TO THE SEA

How do tiny embryos of beach-spawning fishes know when to escape from the confines of the egg and make their way back into the water to begin larval life? An untimely hatch may mean death. Embryos of many species and phyla are aware of their surroundings and respond to cues from the environment (Warkentin, 2011a). Some of these cues will be detailed in a later section, but the most obvious and important cue for fishes incubating on beaches is the presence or absence of water.

In general, even fishes that incubate aquatically will not hatch if out of water (Yamagami, 1988). For beach-spawning fishes at mid to lower tidal heights, submergence occurs daily, and therefore the delay from the time of developmental competence to actual hatching would be only a few hours, at most. This is the case for the kusafugu puffer *Takifugu niphobles* (Yamahira, 1997). In this species of Tetrodontidae, hatching occurs during nocturnal high tides. Over the course of the season, the time of spawning changes slightly so that the time of incubation is sufficient to reach hatching competence when the appropriate tides occur. The specific trigger for hatching has not been investigated in this species, but submergence is required. If the eggs remain emerged into air too long, they will die without hatching.

Other species of fishes that require submergence to hatch include the California grunion *Leuresthes tenuis* (Atherinopsidae) (Griem & Martin, 2000), the rockhopper blenny *Andamia tetradactyla* (Shimizu et al., 2006), the whitebait *Galaxias maculatus* (McDowall, 2006), and some mudskippers (Ishimatsu et al., 2007; Ishimatsu & Graham, 2011). Although this has not been studied extensively, species of fishes that are able to hatch while emerged in air are likely to be rare.

Hatching for kusafugu puffers and California grunion occurs on the flooding tides as waves wash the eggs out of their interstitial hiding places on shore. Once released, the eggs tumble into the surf and hatching occurs aquatically. The eggs of the mummichog *Fundulus heteroclitus* must also be submerged to hatch. This species responds to the flood tide in quiet, still water in an estuary when boundary layers slow oxygen delivery to the embryo (Taylor, 1999). Freed from their chorions, the hatchlings begin the next phase of life as planktonic larvae.

8.3 HATCHING COMPETENCE REQUIRES THE APPROPRIATE STAGE OF DEVELOPMENT

Hatching competence is not the same thing as hatching. An embryo may be developmentally prepared to hatch but not in the proper conditions to initiate the process. Some animals may be competent to hatch before they reach their typical stage for hatching. This frees them to hatch early if necessary to survive in response to predation or some other types of threats (Warkentin, 2011b; Saens et al., 2003), but in the absence of these agents, they continue development to a later stage. Larvae that hatch early may not be as large or as coordinated as larvae that hatch at the usual time, but they survive the attempt at predation, and that is sufficient (Warkentin, 2011a).

For fishes, the stage at hatching competence is highly variable between species. Pelagic eggs have extremely short periods of incubation, and embryos hatch at a very early stage of development, in some cases unable even to swim or feed. In contrast, most beach-spawning fishes have large eggs and long incubation durations that allow the embryos to develop to a more advanced stage, ready to become part of the plankton.

Once hatching competence is reached, hatching usually follows relatively soon. However, in some cases, hatching can be delayed by environmental factors. For example, if the eggs are out of water, they may not hatch. Or if the eggs require an environmental cue, that cue may not be present. In these cases, incubation may be extended for a few hours or a few weeks, depending on the species (Martin, 1999). Amphibian larvae that hatch later than usual may continue to undergo development and outcompete larvae that hatch at the usual time (Sih & Moore, 1993).

8.4 HATCHING IS A TWO-STAGE PROCESS INVOLVING CHEMICAL AND MECHANICAL STEPS

Hatching is a complex two-stage process that involves behavior and physiology (Yamagami, 1988, 1996). The first step is chemical and the next is mechanical. Initiation of hatching requires release of one or more enzymes to digest the chorion (Kawaguchi et al., 2005; Yasumasu et al., 2010). The chorion is made of polymers of glutamate, proline, and glutamine that form cross-linkages during "hardening" after fertilization (Yamagami et al., 1992).

Hatching glands are found on embryos but disappear after hatching occurs. They are typically unicellular glands on the surface of the body, head, or operculum. In many species, but not all, increased activity of the embryo causes the release of the hatching enzyme or chorionase, and hatching follows, usually within 30 minutes. There may be two forms of the chorionase, high and low, to break down the layers of the chorion into peptide fragments from the broken cross-linkages (Yasumasu et al., 2010).

All known chorionases from fishes are highly conserved, as one might expect for such an important molecule (Kawaguchi et al., 2007, 2010). These metalloproteins of the astacin family show similarity to the chorionases of all vertebrate genomes. In teleosts, the high-expression genes for chorionase have evolved with the loss of introns, while the low-expression genes retain their ancestral exon-intron composition (Kawaguchi et al., 2010). Chorionase is inhibited by Ethylenediaminetetraacetic acid (EDTA) (DiMichele et al., 1981; Yamagami, 1996; Martin et al., 2011), indicating calcium or other divalent cations are necessary for its action.

The second stage is the mechanical effort of the embryo to escape from the weakened chorion (Speer-Blank & Martin, 2004). The embryo may spin about within the egg case and expend a good deal of effort and energy as it pushes against the chorion. An anesthetized embryo cannot hatch (Yamagami, 1996; Martin et al., 2011).

8.5 HATCHING MAY BE INITIATED OR SYNCHRONIZED BY ENVIRONMENTAL CUES

Flexibility in hatching takes different forms in different species (Warkentin, 2011a). Species may hatch at different stages of development, earlier or later than normal—but one or the other, rarely both, within a species (Warkentin, 2011b). If age and stage of hatching are fixed or consistent for a given species, the actual initiation of hatching may still respond to presence of aquatic conditions or another environmental cue such as circadian or tidal rhythms within a range of hours in the day. Some species respond to any one of multiple environmental cues (Warkentin & Caldwell, 2009); other species may require a combination of cues presented simultaneously (Griem & Martin, 2000; Martin et al., 2011). The plasticity in hatching has not been studied for most species of beach-spawning fishes, but this group contains many likely candidates for environmentally cued hatching.

There are numerous environmental triggers for hatching. These cues may be mechanical or chemical, too much or too little water, or behavior of parents, siblings, predators, or pathogens (Petranka et al., 1982; Griem & Martin, 2000; Warkentin, 2011a; Ishimatsu et al., 2009; Ishimatsu & Graham, 2011). Many species can alter the timing of hatching in response to an environmental cue, but will hatch on a developmental timetable without the cue. Some species can respond to any of multiple different cues while others seem to respond only to one. Although not all of these cues are known for species of beach-spawning fishes, plasticity of hatching is found across a wide spectrum of organisms from barnacles and flatworms to turtles and frogs (Warkentin, 2011a).

Terrestrial incubation of aquatic eggs certainly provides a strong adaptive impetus for environmentally cued hatching (Martin & Strathmann, 1999; Martin et al., 2004; Warkentin, 2011b), for all the reasons outlined above. However, many aquatic species with aquatic embryos that will become aquatic larvae also rely on cues from the environment to time the moment of hatch, including some invertebrates (Martin & Strathmann, 1999; Strathmann et al., 2010). An environmental cue allows plasticity of response to predators, which may differ between threats to embryos or eggs and threats to swimming larvae. Also, environmentally cued hatching permits synchronization of hatch for a clutch or a cohort (Martin et al., 2004; Baldwin et al., 2004). This sudden influx spreads the risk of entering the plankton among all the members of the cohort and may saturate the predators, allowing the lucky ones to find their way into their new environment.

The effectiveness of environmental cues depends on the ability of the embryos to perceive the sensory information and integrate it for purposeful activity. Thus an appropriate stage of development must be reached and the hatching enzyme must be produced by the hatching glands. Once hatching competence is reached, there may be a window of time when hatching can occur if the cue is present. At some point, if the embryo does not emerge, it becomes unable to respond to the cue, either because of lack of energy reserves or for some other reason (Martin, 1999; Smyder & Martin, 2002).

Environmental cues and hatching have been examined in three beach-spawning fishes, the mummichog *Fundulus heteroclitus* that spawns on vegetation and shells along the edge of estuaries, the California grunion *Leuresthes tenuis* that spawns on sandy beaches, and the mudskipper *Periophthalmus modestus* that lives and reproduces in burrows on exposed mud flats. Each is uniquely suited to its habitat, and each is described in detail in the following sections.

8.6 THE MUMMICHOG *FUNDULUS HETEROCLITUS* HATCHES IN RESPONSE TO A CHEMICAL CUE, AQUATIC HYPOXIA

The mummichog *F. heteroclitus* is a model species for study of reproduction, hatching, and genetics. Its eggs are spawned over biogenic substrates such as bivalve shells or vegetation high in the intertidal zone during a semilunar flood tide. When the tide recedes, the eggs incubate emerged into air but shaded and protected from desiccation by their spawning substrate. They develop to the point of competence to hatch within the two weeks between semilunar tides. The embryos will not hatch while emerged in air, but must be submerged in brackish water or seawater in order to hatch (DiMichele & Taylor, 1980, 1981). If the tides do not reach them, they can remain unhatched but alive for additional days or weeks, but the development rate slows and they are vulnerable to predation and disease (DiMichele & Powers, 1991).

The environmental cue for hatching is the aquatic hypoxia that develops around the eggs when they are submerged. This forms because of boundary layers of water that are depleted of oxygen by the actively metabolizing embryos. The increased gill ventilation rate triggers the release of chorionase from the hatching glands in the operculum, and within about 30 minutes the embryo begins to hatch out of the egg (Taylor & DiMichele, 1983), having released its chorionase from the operculum and oral margin (DiMichele et al., 1981). The highly conserved chorionases for this species have been sequenced for comparison with other vertebrates (Kawaguchi et al., 2005, 2010).

Significant differences between populations in hatching time and other aspects of development have a genetic basis in this species (DiMichele & Powers, 1991; Whitehead et al., 2011). Aquatic hypoxia was one of the first environmental cues described for hatching and it is also seen in many species of amphibians (Petranka et al., 1982; Gunzburger, 2003; Warkentin, 2011b).

A false cue to hatch may happen when rainfall surrounds the eggs; unfortunately, if these estuarine hatchlings emerge into freshwater, they soon die (Williamson & DiMichele, 1997). But before the embryos are competent to hatch, rainfall cannot trigger this response. This suggests that balancing selection is operating on development rate for *Fundulus heteroclitus* to see that they don't hatch too soon with false cues but are ready at the correct time when the right cue occurs (Williamson & DiMichele, 1997). Hatching success is best at salinities between 10 and 30 parts per thousand, or ppt (normal seawater is 30–35 ppt), but some hatching can occur even in freshwater and double-strength seawater (Tay & Garside, 1975).

8.7 THE CALIFORNIA GRUNION *LEURESTHES TENUIS* HATCHES IN RESPONSE TO A MECHANICAL CUE, AGITATION IN SEAWATER

The California grunion *Leuresthes tenuis* spawns completely out of water, and its eggs also incubate completely out of water (Walker, 1952). It has long been known that the embryos hatch out when they are washed into the ocean at the following semilunar high tide (Thompson, 1919). The actual process of hatching is much more rapid than that seen in most fishes, including the environmentally cued hatching of *F. heteroclitus* described above (Martin et al., 2011). Unlike the estuarine habitat of the mummichog, the grunion lives on the outer coast on sandy beaches that are pounded by surf waves. When the tides rise and waves wash out the egg-encased embryos, they enter water that is normoxic or even hyperoxic, full of air bubbles (Griem & Martin, 2000).

Careful testing of the embryos of California grunion can tease apart the specific cues that embryos use for hatching. Embryos that are competent to hatch and placed carefully into still seawater do not hatch (David, 1939; Griem & Martin, 2000). Grunion embryos that are shaken but not immersed in water do not hatch. Placed in freshwater and shaken vigorously, embryos die without hatching out of the eggs. Embryos that are held in place but surrounded by swirling water do not hatch (Martin et al., 2011). The necessary requirements for hatching in *L. tenuis* are agitation of the eggs in seawater; these are provided by waves pounding on the sandy shore in nature. Under these conditions hatching occurs within one or two minutes (Griem & Martin, 2000). Changes in salinity affect development and hatching (Matsumoto & Martin, 2008), as do changes in the ionic composition of artificial seawater (Martin et al., 2011).

Low level of oxygen is not the trigger for hatching in *L. tenuis*; in fact, aquatic hypoxia slows hatching down (Griem & Martin, 2000). High oxygen levels inhibit hatching in *F. heteroclitus* but make hatching more rapid in *L. tenuis*. The environmental cues that trigger hatching are compared for *L. tenuis* and *F. heteroclitus* in Table 8.1.

TABLE 8.1

Comparison of Hatching Cues for Two Fishes with Terrestrial Incubation and Environmentally Cued Hatching, *Leuresthes tenuis* and *Fundulus heteroclitus*, and One Fish That Incubates in Water, *Oryzias latipes*

	L. tenuis (marine)	F. heteroclitus (estuarine)	O. latipes (freshwater)
Air Emergence	Inhibits	Inhibits	Inhibits
Hyperoxic water	Permissive	Inhibits	Inhibits
Still H$_2$O, low oxygen	Permissive	Stimulates	Stimulates
Still H$_2$O, normal oxygen	Permissive	Inhibits	Inhibits
Agitated H$_2$O, high oxygen	Stimulates	Inhibits	Inhibits
Agitated, hypoxic water	Stimulates	Stimulates	Stimulates
Hatching glands	Lateral surface of body	Pharyngeal cavity and periphery of mouth	Pharyngeal cavity and periphery of mouth
Speed of hatch	0.5 to 5 min	30 to 60 min	30 to 60 min
MS-222	Inhibits	Inhibits	Inhibits

If the cue to hatch is not present, embryos of *L. tenuis* can extend incubation up to three times longer than the time to reach primary hatching competence (Darken et al., 1998; Smyder & Martin, 2002). The embryos remain alert and can respond by hatching instantly if the cue is presented at any time (Speer-Blank & Martin, 2004). The development during this period is arrested for most parts of the body; see Section 8.7 for additional information about extended incubation.

It is very likely that mechanical agitation cues hatching in other species of fishes, including some others that spawn on beaches. The terrestrially incubating red-eyed tree frog also hatches in response to mechanical stimulation (Warkentin, 1995). There is some evidence that this environmental cue is important for embryos across a range of taxa; this intriguing possibility deserves further study.

8.8 THE MUDSKIPPER *PERIOPHTHALMUS MODESTUS* HATCHES IN RESPONSE TO PARENTAL INTERVENTION AND AQUATIC HYPOXIA

Mudskippers are highly amphibious fishes that live on mud flats in estuaries of tropical and temperate regions of the Indo-Pacific (Clayton, 1993). They are very active terrestrially during low tides and construct elaborate burrows into the mud for shelter. Water within the burrow becomes hypoxic because of contact with the anoxic burrow mud and because it may be stagnant during a low tide that may last weeks during some seasons. In the burrows of some species of mudskippers are chambers that are open underneath and connect with a main burrow by a J-shaped tunnel. These chambers can be filled with air when the fish takes a gulp of air in its mouth, seals the operculae, and swims into the burrow to release it in the top of the chamber (Ishimatsu et al., 1998). With multiple mouthfuls, the parent creates an air-filled chamber under the mud, protected from sunlight and desiccation. Here, the embryos are protected from hypoxia that would be deadly, by being emerged above the water (Etou et al., 2007).

Air chambers may be created and filled by adult fishes at various times during the year. At least some species, including *Periophthalmus modestus*, nest within these chambers in burrows. Spawning has not been observed but may occur in the air for *Periophthalmodon schlosseri* (Ishimatsu et al., 2009). When the eggs are laid, they are placed in a monolayer on the ceiling of the chamber, so that the air phase allows them to be out of touch with the hypoxic water that would otherwise surround them. The guarding parents periodically replace and replenish the air with fresh mouthfuls.

When the embryos are competent to hatch, a flood tide cues the guarding parent to remove the air. He rapidly shuttles back and forth, taking mouthful after mouthful of air from the chamber to the foot of the main tunnel and releasing it, where it can rise to the surface unimpeded. This action submerges the clutch of eggs in hypoxic water, and they hatch rapidly (Ishimatsu et al., 2007). Submergence of the burrow and the surrounding area of the mud flats by the incoming tide allows the larvae to exit the burrow and swim free of their natal home. The eggs must be submerged to hatch, and the parental behavior ensures aerial incubation followed by aquatic submergence at the proper times, a unique solution to the problem of embryonic development in the mud (Ishimatsu & Graham, 2011).

8.9 EXTENDED INCUBATION MAY ARREST DEVELOPMENT BUT DIFFERS FROM METABOLIC ARREST DURING DIAPAUSE

Three major forms of arrested development have been described for fish embryos: embryonic diapause, anoxia-induced quiescence, and delayed hatching (Podrabsky et al., 2010).

Embryonic diapause, the most dramatic form of arrested development, is seen in annual killifishes that live in seasonal pools in the tropics of South America and Africa. In *Austrofundulus limnaeus*, this (usually) obligatory period of dormancy stops development temporarily but completely, and metabolism is slowed to the point that cellular activity is almost undetectable (Wourms, 1972; Podrabsky & Hand, 1999). Embryonic diapause occurs in three phases during drought when the embryos are buried in hypoxic mud. During each phase the metabolism is extremely low and the embryo is unresponsive to stimuli.

Anoxia-induced quiescence is a period of dormancy that is initiated by extremely low availability of oxygen. Prolonged anoxia is usually fatal for most embryos, but some annual killifishes have adapted to enter dormancy under these conditions, and they survive (Podrabsky et al., 2010). This is a response to environmental conditions, rather than a set developmental program as for embryonic diapause, although both are characterized by extremely low metabolic rates. Some adult fishes are able to survive winter with a period of torpor brought on by a combination of cold temperatures and anoxia. Turtles lay eggs on land like most reptiles, but if eggs of *Chelodina rugosa* are submerged in water before they are ready to hatch, they can enter an arrested state and survive for hours or days because the cells are not consuming oxygen (Kennett et al., 1993).

Delayed hatching occurs after embryonic development is complete and hatching competence is achieved (Martin, 1999; Moravek & Martin, 2011). Most beach-spawning fishes are able to delay hatching for a few hours in order to wait for submergence by tides (Yamahira, 1996). This delay in hatching is brief and may not be unique to terrestrially incubating fishes (Yamagami, 1988). On the other hand, some species are able to extend incubation for days or even weeks and still hatch successfully to begin larval life (Moffatt & Thomson, 1978; Taylor, 1999; Smyder & Martin, 1999).

Some fishes and amphibians that incubate eggs terrestrially and have environmentally cued hatching can delay hatching long past the time of hatching competence (Martin, 1999; Warkentin, 2011b; Martin & Carter, 2013). This is seen in the terrestrial eggs of some beach-spawning fishes in the families Fundulidae, Atherinidae, and Galaxiidae. It has been well studied for two species, *Fundulus heteroclitus* and *Leuresthes tenuis*. *Galaxias masculatus* can extend incubation for up to six weeks post fertilization after reaching hatching competence in 14 days, but little is known about this (McDowall, 2006).

For fishes with terrestrial embryos, eggs on a beach that do not hatch during one semilunar tidal cycle are able to wait for a couple of weeks until the next semilunar high tides before setting out to sea (Taylor et al., 1977; Darken et al., 1998). Once hatching competence has been attained, whenever the cue arrives, hatching can occur at any time (DiMichele & Taylor, 1981; Griem & Martin, 2000; Speer-Blank & Martin, 2004; Martin et al., 2011). However, if the environmental cue to hatch does not arrive, then the embryo eventually will die without hatching (Figure 8.1).

The mummichog *Fundulus heteroclitus* can survive as an embryo up to 37 days post fertilization (dpf), after being ready to hatch around 9 dpf (DiMichele & Taylor, 1981). Potential environmental cues that have been tested include salinity, pH, dissolved oxygen, hydration, and temperature (DiMichele & Taylor, 1980). This hardy estuarine species has a broad salinity and temperature tolerance, and physical conditions of salinity, pH, and temperature within its adult tolerance did not alter embryo survival or hatching. Eggs of *F. heteroclitus* do not hatch while emerged into air and hatch more slowly after submersion if they were previously dehydrated. Submersion in normoxic water does not trigger hatching, nor does aerating or agitating the eggs in water. For this species, aquatic hypoxia is necessary to trigger hatching, a condition that happens in nature with still waters and flood tides in the estuary (DiMichele & Taylor, 1980, 1981). The continuation or cessation of development has not been examined in detail for this species, but metabolism is not depressed for up to 14 dpf, a week after hatching competence is achieved (Taylor et al., 1977), and oxygen consumption and opercular ventilation increase during hatching (DiMichele & Powers, 1984).

Although the developmental stages are the same in air as for *F. heteroclitus* embryos in water (Tingaud-Sequeira et al., 2009), over 2380 putative genes are expressed differently during aerial incubation (Tingaud-Sequeira et al., 2013).

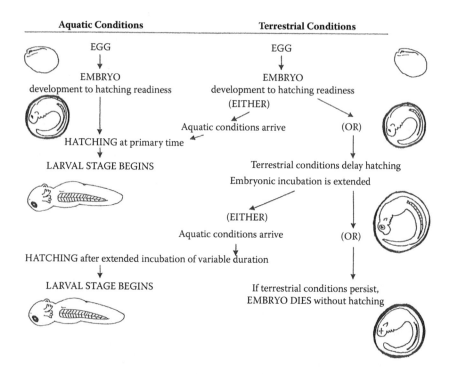

FIGURE 8.1   Environmentally cued hatching can result in hatching at the time of earliest competence if the cue is present. If absent, the incubation can be prolonged until the cue appears. If the cue fails to appear, the embryo may die without hatching. (From Martin, K. L. M., *American Zoologist*, 39, 1999, 279–288. Art by Gregory Martin.)

Some of these proteins are also seen in "stress" pathways and may be protecting the metabolically active embryos from harm from reactive oxygen species, or providing molecular chaperones, adding to the effect of downregulation of aquaporin channels, that happens within hours of air exposure of *F. heteroclitus* eggs.

For the California grunion *Leuresthes tenuis*, incubation can be extended up to 40 dpf, although hatching competence occurs at about 9 or 10 dpf (Martin et al., 2009; Moravek & Martin, 2011). The delay of hatching is not required; if the environmental cue to hatch arrives, embryos can hatch as soon as they are developmentally capable. Eggs of *L. tenuis* that incubate in aerated seawater, agitated with the current, hatch spontaneously at around 9 to 10 dpf at 20°C (David, 1939). During extended incubation for *L. tenuis*, metabolism continues at a steady rate (Darken et al., 1998). This results in a temperature-specific decline in yolk energy stores until a point is reached that the embryo can no longer hatch (Smyder & Martin, 2002).

Hatchlings must begin feeding within four days or they will die (May, 1971), but within the egg, the embryo can survive much longer, even weeks after hatching competence. The metabolic rate increases during hatching (Martin et al., 2011), and of course larvae lead active swimming lives as well as being busy developing their bodily structures, so energy needs are higher after hatching than before.

Because of this flexibility in hatching time, a cohort of embryos from the same spawning date can be hatched at different numbers of days post fertilization, and then followed for larval growth (Martin et al., 2010). Hatchlings from the primary hatching period, around 10 dpf, grow rapidly and feed immediately. Larval development of fins and internal organs commences (Moravek & Martin, 2011). If additional cohorts from the same spawning date are hatched after 20 or 30 days of incubation rather than 10, the delayed hatchlings are similar in length to hatchlings of the primary hatching time. Once hatched, they grow rapidly at the same rate as those larvae from the primary hatch period, but the development and size of the newly hatched larvae on any given date is not at the same stage as the development and size of those in the cohort that hatched earlier. Figure 8.2 shows the growth of larvae with different hatch dates but the same day of fertilization, and the growth of a hatched larva compared to the growth of an embryo with the same fertilization date.

When hatching competence is achieved by embryos of *L. tenuis*, most development ceases until after hatching occurs (Moravek & Martin, 2011). During the extended incubation period following hatching competence but preceding actual hatch, there is no change in length of the embryo, and only very subtle changes in appearance. Melanophores do not increase in number but become more heavily pigmented, the otoliths grow in diameter, and a small number of teeth develop (Figure 8.3). These structures that continue slowly growing are all neural crest cell derivatives. Otoliths put down daily increments in *L. tenuis* during larval life as well (Brothers et al., 1976).

The way that *L. tenuis* arrest development while maintaining metabolic activity is different from the way some other amphibians respond to delayed hatching.

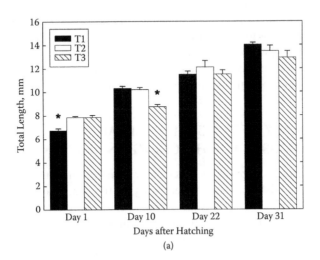

FIGURE 8.2   (a) Growth after hatching is similar for a cohort of *Leuresthes tenuis*, the California grunion, from the same fertilization date but different hatching dates. T1 larvae were hatched 10 days post fertilization (dpf), T2 at 20 dpf, and T3 at 30 dpf. (From Martin, K. L., Moravek, C. L., and Walker, A. J., *Environmental Biology of Fishes*, 2010, DOI: 10.1007/s10641-010-9760-4.)

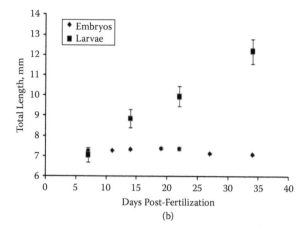

FIGURE 8.2 *(Continued)* (b) The growth of two cohorts with the same day of fertilization differs when one group extends incubation and remains embryonic, while the second is triggered to hatch and begin larval life.

The Australian toadlet *Pseudophryne bibroni* continues slow development during an incubation period that may be extended for months waiting for rains to return and trigger hatching (Bradford & Seymour, 1985). The salamander *Amphiuma means* lays its eggs on the edge of ponds that dry during the incubation period, and development continues to the point that when the rains return, some hatchlings have already achieved adult morphology (Gunzburger, 2003).

The rate of development changes upon hatching; the instant that hatching occurs, larval life begins. This form of plasticity of development is called *heterokairy* (Spicer & Burggren, 2003), when individuals of the same species at the same age can be at very different stages of development because of environmental conditions. The regulatory control of development in this species deserves careful study.

8.10 SUMMARY OF THE CHALLENGES AND MECHANISMS FOR HATCHING ON BEACHES

Finding the helpless embryos of fishes on dry land prompts investigation of how they will accomplish the task of returning to the sea. Many beach-spawning parent fishes disappear back into the ocean once the oviposition and fertilization are done. The extreme danger of hatching as an aquatic larva into an unforgiving terrestrial environment is prevented by environmentally cued hatching, even if this results simply from a lack of efficacy or inability to secrete the chorionase when the egg is out of water. Synchronization of hatching by tidal cues or parental behavior may help embryos to survive by overwhelming predators with large numbers entering the plankton at once. If extended incubation is possible, embryos of the same fertilization date may hatch at different times and develop through the larval stages asynchronously. Table 8.2 shows examples of the different types of

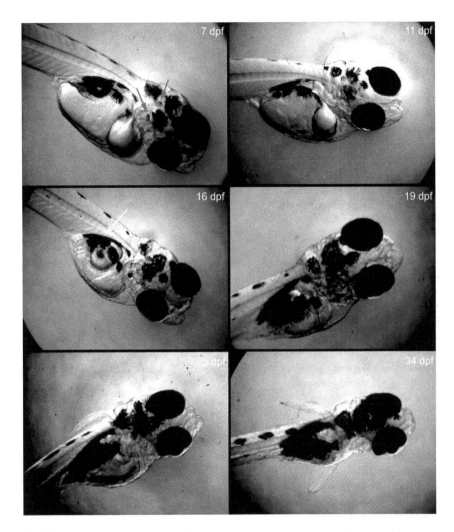

FIGURE 8.3 Morphology of hatchlings does not change significantly even when incubation time extends long after the time of hatching competence. Days post fertilization (dpf) are shown for each. California grunion (*Leuresthes tenuis*) embryos are competent to hatch at about 10 dpf. Arrows indicate otoliths (7 dpf, 25 dpf), head melanophores (11 dpf), and gut melanophores (16 dpf). (From Moravek, C. L., and Martin, K. L., *Copeia*, 2011, 308–314, DOI: 10.1643/CG-10-164.)

timing for hatching and embryonic development in beach-spawning fishes and terrestrially breeding amphibians (Martin & Carter, 2013).

Embryonic hatching plasticity may take many forms and allow early or late hatching in response to cues or threats in the environment. In many areas of the coast, anthropogenic activities or alterations of the shoreline pose new threats to the successful incubation of species of fishes that spawn on beaches. Chapter 9 discusses some of the threats facing beach-spawning fishes today.

TABLE 8.2
Types of Hatching for Beach-Spawning Fishes and Terrestrially Nesting Amphibians, with Some Examples

Early Development in Terrestrial Habitat	Example Species	Reference
Conservative Constitutive: Continuous development to hatching competence, hatch even if emerged, no environmental cue for hatching	Capelin *Mallotus villosus*	Frank & Leggett, 1981
ECH Early Alert: Continuous development to hatching competence, hatch even if emerged, potential to hatch early if disturbed by predators or other threats	Red-eyed tree frog *Agalychnis callidryas*	Warkentin, 2011
Cautious Constitutive: Continuous development to hatching competence, hatch only upon tidal immersion	Kusafugu puffer *Takifugu niphobles*	Yamahira, 1996
ECH with Parental Involvement: Continuous development to hatching competence, hatch upon immersion caused by parental behavior	Mudskipper *Periophthalmus modestus*	Ishimatsu & Graham, 2011
ECH Ready and Waiting: Continuous development to hatching competence, potential for prolonging incubation duration with active metabolism but developmental stasis, environmentally cued hatching	California grunion *Leuresthes tenuis*	Moravek & Martin, 2011
ECH Ready and Progressing: Continuous development to hatching competence, potential for prolonged incubation duration while slowly continuing developmental progress, environmentally cued hatching	Australian toadlet *Pseudophryne bibroni*	Bradford & Seymour, 1985
Precocious: Continuous development beyond usual hatching stage, prolonged incubation while continuing developmental progress, hatch as juvenile completely bypassing aquatic larval stage	Foam nesting frogs, *Leptocatylus fuscus*	deCarvalho et al., 2012
Diapause: Discontinuous development with obligatory periods of metabolic and maturational stasis, environmentally cued hatching	Annual killifish, *Austrofundulus limnaeus*	Podrabsky et al., 2010

Source: Martin, K. L., and Carter, A. L., *Integrative and Comparative Biology*, 53, 233–247, 2013.

REFERENCES

Altig, R. & McDiarmid, R. W. (2010). Morphological diversity and evolution of egg and clutch structure in amphibians. *Herpetological Monographs* 21, 1–32.

Baldwin, J. L., Goldsmith, C. E., Petersen, C. E., Preston, R. L. & Kidder, G. W. (2004). Synchronous hatching in the *Fundulus heteroclitus* embryos: Production and properties. *Bulletin of the Mount Desert Island Biological Laboratory* 43, 110–111.

Bradford, D. F. & Seymour, R. S. (1985). Energy conservation during the delayed hatching period in the frog *Pseudophryne bibroni*. *Physiological Zoology* 58, 491–456.

Brothers, E. B., Mathews, C. P. & Lasker, R. (1976). Daily growth increments in otoliths from larval and adult fishes. *Fishery Bulletin* 74, 1–8.

Clayton, D. A. (1993). Mudskippers. *Oceanography and Marine Biology Annual Review* 31, 507–577.

Darken, R. S., Martin, K. L. M. & Fisher, M. C. (1998). Metabolism during delayed hatching in terrestrial eggs of a marine fish, the grunion *Leuresthes tenuis*. *Physiological Zoology* 71, 400–406.

David, L. R. (1939). Embryonic and early larval stages of the grunion, *Leuresthes tenuis*, and of the sculpin, *Scorpaena guttata*. *Copeia* 1939, 75–80.

DiMichele, L. & Powers, D. A. (1984). The relationship between oxygen consumption rate and hatching in *Fundulus heteroclitus*. *Physiological Zoology* 57, 46–51.

DiMichele, L. & Powers, D. A. (1991). Allozyme variation, developmental rate, and differential mortality in the teleost *Fundulus heteroclitus*. *Physiological Zoology* 64, 1426–1443.

DiMichele, L. & Taylor, M. (1980). The environmental control of hatching in *Fundulus heteroclitus*. *Journal of Experimental Zoology* 214, 181–187.

DiMichele, L. & Taylor, M. (1981). The mechanism of hatching in *Fundulus heteroclitus*: Development and physiology. *Journal of Experimental Zoology* 217, 73–79.

DiMichele, L., Taylor, M. H. & Singleton, R. Jr. (1981). The hatching enzyme of *Fundulus heteroclitus*. *Journal of Experimental Zoology* 216, 133–140.

Etou, A., Takeda, T., Yoshida, Y. & Ishimatsu, A. (2007). Oxygen consumption during embryonic development of the mudskipper (*Periophthalmus modestus*): implication for the aerial development in burrows. In *Fish Respiration and Environment* (Fernandes, M. N., Rantin, F. T., Glass, M. L. & Kapoor, B. G., eds.), 83–91. Enfield, NJ: Science Publishers.

Frank, K. T. & Leggett, W. C. (1981a). Wind regulation of emergence times and early larval survival in capelin (*Mallotus villosus*). *Canadian Journal of Fisheries and Aquatic Science* 38, 215–223.

Frank, K. T. & Leggett, W. C. (1981b). Prediction of egg development and mortality rates in capelin (*Mallotus villosus*) from meteorological, hydrographic, and biological factors. *Canadian Journal of Fisheries and Aquatic Science* 38, 1327–1338.

Geiser, F. & Seymour, R. S. (1989). Influence of temperature and water potential on survival of hatched, terrestrial larvae of the frog *Pseudophryne bibronii*. *Copeia* 1989, 207–209.

Griem, J. N. & Martin, K. L. M. (2000). Wave action: the environmental trigger for hatching in the California grunion *Leuresthes tenuis* (Teleostei: Atherinopsidae). *Marine Biology* 137, 177–181.

Gunzburger, M. S. (2003). Evaluation of the hatching trigger and larval ecology of the salamander *Amphiuma means*. *Herpetologica* 59, 459–469.

Heming, T. A. & Buddington, R. K. (1988). Yolk absorption in embryonic and larval fishes. In *Fish Physiology Vol. 11 A* (Hoar, W. S. & Randall, D. J., eds.), 407–446. New York: Academic Press.

Ishimatsu, A. & Graham, J. B. (2011). Roles of environmental cues for embryonic incubation and hatching in mudskippers. *Integrative and Comparative Biology* 51, 38–48. DOI: 10.1093/icb/icr018.

Ishimatsu, A., Takeda, T., Tsuhako, Y., Gonzales, T. T. & Khoo, K. H. (2009). Direct evidence for aerial egg deposition in the burrows of the Malaysian mudskipper *Periophthalmodon schlosseri*. *Ichthyological Research* 56, 417–420.

Ishimatsu, A., Yoshida, Y., Itoki, N., Takeda, T., Lee, H. J. & Graham, J. B. (2007). Mudskippers brood their eggs in air but submerge them for hatching. *Journal of Experimental Biology* 210, 3946–3954.

Ishimatsu, A., Hishida, Y., Takita, T., Kanda, T., Oidawa, S., Takeda, T. & Khoo, K. H. (1998). Mudskippers store air in their burrows. *Nature* 391, 237–238.

Kawaguchi, M., Yasumasu, S., Hiroi, J., Naruse, K., Suzuki, T. & Iuchi, I. (2007). Analysis of the exon-intron structures of fish, amphibian, bird, and mammalian hatching enzyme genes, with special reference to the intron loss evolution of hatching enzyme genes in Teleostei. *Gene* 392, 77–88. DOI: 10.1016/j.gene.2006.11.012.

Kawaguchi, M., Yasumasu, S., Shimizu, A., Hiroi, J., Yoshizaki, N., Nagata, K., Tanokura, M. & Iuchi, I. (2005). Purification and gene cloning of *Fundulus heteroclitus* hatching enzyme. A hatching enzyme system composed of high choriolytic enzyme and low choriolytic enzyme is conserved between two different teleosts, *Fundulus heteroclitus* and medaka *Oryzias latipes*. *FEBS Journal* 272, 4315–4326.

Kawaguchi, M., Yasumasu, S., Shimizu, A., Sano, K., Iuchi, I. & Nishida, M. (2010). Conservation of the egg envelope digestion mechanism of hatching enzyme in euteleostean fishes. *FEBS Journal* 277, 4973–4987.

Kennett, R., Georges, A. & Palmer-Allen, M. (1993). Early developmental arrest during immersion of eggs of a tropical freshwater turtle, *Chelodina rugosa* (Testudinata: Chelidae), from northern Australia. *Australian Journal of Zoology* 41, 37–45.

Martin, K. L. M. (1999). Ready and waiting: delayed hatching and extended incubation of anamniotic vertebrate terrestrial eggs. *American Zoologist* 39, 279–288.

Martin, K. L. & Carter, A. L. (2013). Brave new propagules: terrestrial embryos in anamniotic eggs. *Integrative and Comparative Biology* 53, 233–247. DOI:10.1093/icb/ict018.

Martin, K. L. M. & Strathmann, R. (1999). Aquatic organisms, terrestrial eggs: early development at the water's edge. Introduction to the symposium. *American Zoologist* 39, 215–217.

Martin, K. L. M., Bailey, K., Moravek, C. & Carlson, K. (2011). Taking the plunge: California grunion embryos emerge rapidly with environmentally cued hatching. *Integrative and Comparative Biology* 51, 26–37. DOI: 10.1093/icb/icr037.

Martin, K. L., Moravek, C. L. & Flannery, J. A. (2009). Embryonic staging series for the beach spawning, terrestrially incubating California grunion *Leuresthes tenuis* (Ayres 1860) with comparisons to other Atherinomorpha. *Journal of Fish Biology* 75, 17–38.

Martin, K. L., Moravek, C. L. & Walker, A. J. (2010). Waiting for a sign: extended incubation postpones larval stage in the beach spawning California grunion *Leuresthes tenuis* (Ayres). *Environmental Biology of Fishes* 91, 63–70. DOI: /10.1007/s10641-010-9760-4.

Martin, K. L. M., Van Winkle, R. C., Drais, J. E. & Lakisic, H. (2004). Beach spawning fishes, terrestrial eggs, and air breathing. *Physiological and Biochemical Zoology* 77, 750–759.

Matsumoto, J. K. & Martin, K. L. M. (2008). Lethal and sublethal effects of altered sand salinity on embryos of beach-spawning California grunion. *Copeia* 2008, 483–490.

May, R. C. (1971). Effects of delayed initial feeding on larvae of the grunion, *Leuresthes tenius* (Ayers). *U.S. Fisheries Bulletin* 69, 411–425.

McDowall, R. M. (2006). Crying wolf, crying foul, or crying shame: alien salmonids and a biodiversity crisis in the southern cool-temperate galaxoid fishes? *Reviews in Fish Biology and Fisheries* 16, 233–422. DOI: 10.1007/s11160-006-9017-7.

Moffatt, N. M. & Thomson, D. A. (1978). Tidal influence on the evolution of egg size in the grunions (*Leuresthes*, Atherinidae). *Environmental Biology of Fishes* 3, 267–273.

Moravek, C. L. & Martin, K. L. (2011). Life goes on: delayed hatching, extended incubation, and heterokairy in development of embryonic California grunion, *Leuresthes tenuis*. *Copeia* 2011, 308–314. DOI: 10.1643/CG-10-164.

Nakashima, B. S. & Taggart, C. T. (2002). Is beach-spawning success for capelin, *Mallotus villosus* (Muller), a function of the beach? *ICES Journal of Marine Science* 59, 897–908.

Nakashima, B. S. & Wheeler, J. P. (2002). Capelin (*Mallotus villosus*) spawning behavior in Newfoundland waters: the interaction between beach and demersal spawning. *ICES Journal of Marine Science* 59, 909–916.

Petranka, J. W., Just, J. J. & Crawford, E. C. (1982). Hatching of amphibian eggs: the physiological trigger. *Science* 217, 257–259.

Podrabsky, J. & Hand, S. (1999). The bioenergetics of embryonic diapauses in an annual killifish, *Austrofundulus limnaeus*. *Journal of Experimental Biology* 202, 2567–2580.

Podrabsky, J. E., Tingaud-Sequeira, A. & Cerdà, J. (2010). Metabolic dormancy and responses to environmental desiccation in fish embryos. In *Topics in Current Genetics: Dormancy and Resistance in Harsh Environments,* Volume 21 (Lubzens, E., Cerdà, J., & Clark, M., eds.), 203–226. Heidelberg, Dordrecht, London, New York: Springer.

Saens, D., Johnson, J. B., Adams, C. K. & Dayton, G. H. (2003). Accelerated hatching of southern leopard frog (*Rana sphenocephala*) eggs in response to the presence of a crayfish (*Procambarus nigrocinctus*) predator. *Copeia* 2003, 646–649.

Sih, A. & Moore, R. D. (1993). Delayed hatching of salamander eggs in response to enhanced larval predation risk. *American Naturalist* 142, 947–960.

Smyder, E. A. & Martin, K. L. M. (2002). Temperature effects on egg survival and hatching during the extended incubation period of the California grunion, *Leuresthes tenuis*. *Copeia* 2002, 313–320.

Speer-Blank, T. & Martin, K. L. M. (2004). Hatching events in the California grunion, *Leuresthes tenuis*. *Copeia* 2004, 21–27.

Spicer, J. I. & Burggren, W. W. (2003). Development of physiological regulatory systems: altering the timing of crucial elements. *Zoology* 106, 91–99.

Strathmann, R. R., Strathmann, M. F., Ruiz-Jones, G. & Hadfield, M. G. (2010). Effect of plasticity in hatching on duration as a precompetent swimming larva in the nudibranch *Phestilla sibogae*. *Invertebrate Biology* 129, 309–318.

Tay, K. L. & Garside, E. T. (1975). Some embryogenic responses of mummichog, *Fundulus heteroclitus* (L.) (Cyprinodontidae), to continuous incubation in various combinations of temperature and salinity. *Canadian Journal of Zoology* 53, 920–933.

Taylor, M. H. (1999). A suite of adaptations for intertidal spawning. *American Zoologist* 39, 313–320.

Taylor, M. H. & DiMichele, L. (1983). Spawning site utilization in a Delaware population of *Fundulus heteroclitus* (Pisces: Cyprinodontidae). *Copeia* 1983, 719–725.

Taylor, M. H., DiMichele, L., & Leach, G. J. (1977). Egg stranding in the life cycle of the mummichog, *Fundulus heteroclitus*. *Copeia* 1977, 397–399.

Thompson, W. F. (1919). The spawning of the grunion (*Leuresthes tenuis*). *California Fish and Game* 5, 1–27.

Tingaud-Sequeira, A., Lozano, J.-J., Zapater, C., Otero, D., Kube, M., Reinhardt, R. & Cerda, J. (2013). A rapid transcriptome response is associated with desiccation resistance in aerially-exposed killifish embryos. *PLOS One*, 8, 5, e64410.

Tingaud-Sequeira, A., Zapater, C., Chauvigné, F., Otero, D. & Cerdà, J. (2009). Adaptive plasticity of killifish (*Fundulus heteroclitus*) embryos: dehydration-stimulated development and differential aquaporin-3 expression. *American Journal of Physiology: Regulatory, Integrative, and Comparative Physiology* 296, 1041–1052.

Walker, B. W. (1952). A guide to the grunion. *California Fish and Game* 3, 409–420.

Warkentin, K. (1995). Adaptive plasticity in hatching age: A response to predation risk trade-offs. *Proceedings of the National Academy of Sciences, U.S.A.* 92, 3507–3510.

Warkentin, K. M. (2011a). Environmentally cued hatching across taxa: embryos respond to risk and opportunity. *Integrative and Comparative Biology* 51, 14–25. DOI: 10.1093/icb/icr017.

Warkentin, K. M. (2011b). Plasticity of hatching in amphibians: evolution, trade-offs, cues, and mechanisms. *Integrative and Comparative Biology* 51, 111–127. DOI: 10.1093/icb/icr046.

Warkentin, K. M. & Caldwell, M. S. (2009). Assessing risk: embryos, information, and escape hatching. In *Cognitive Ecology II* (Dukas R., & Ratcliffe, J. M. eds.), 177–200. Chicago: University of Chicago Press.

Whitehead, A., Galvez, F., Zhang, S., Williams, L. M. & Oleksiak, M. F. (2011). Functional genomics of physiological plasticity and local adaptation in killifish. *Journal of Heredity* 102, 499–511.

Williamson, E. G. & DiMichele, L. (1997). An ecological simulation reveals balancing selection acting on development rate in the teleost *Fundulus heteroclitus. Marine Biology* 128, 9–15.

Wourms, J. P. (1972). The developmental biology of annual fishes. III. Pre-embryonic and embryonic diapause of variable duration in the eggs of annual fishes. *Journal of Experimental Zoology* 182, 389–414.

Yamagami, K. (1988). Mechanisms of hatching in fish. *Fish Physiology, Volume 11A* (Hoar, W. S. & Randall, D. J., eds.), 447–499. San Diego, CA: Academic Press.

Yamagami, K. (1996). Studies on the hatching enzyme (choriolysin) and its substrate, egg envelope, constructed of the precursors (choriogenins) in *Oryzias latipes*: a sequel to the information in 1991/1992. *Zoological Science* 13, 331–340.

Yamagami, K., Hamazaki, T. S., Yasumasu, S., Masuda, K. & Iuchi, I. (1992). Molecular and cellular basis of formation, hardening, and breakdown of the egg envelope in fish. *International Reviews in Cytology* 136, 51–92.

Yamahira, K. (1996). The role of intertidal egg deposition on survival of the puffer, *Takifugu niphobles* (Jordan et Snyder), embryos. *Journal of Experimental Marine Biology and Ecology* 198, 291–306.

Yamahira, K. (1997). Hatching success affects the timing of spawning by the intertidally spawning puffer *Takifugu niphobles. Marine Ecology Progress Series* 155, 239–248.

Yasumasu, S., Kawaguchi, M., Ouchi, S., Sano, K., Murata, K., Sugiyama, H., Akema, T. & Iuchi, I. (2010). Mechanism of egg envelope digestion by hatching enzymes, HCE and LCE in medaka, *Oryzias latipes. Journal of Biochemistry* 148, 439–448.

9 Coastal Squeeze: New Threats to Beach-Spawning Fishes and Their Critical Habitats

Spawning intertidally is physiologically demanding, but may provide a release from predation pressure. However, it opens up a whole new world of impacts from human developments.

—Dan Penttila (2013, personal communication)

Beaches are coastal ecosystems that are affected by both marine and terrestrial influences (McLachlan & Brown, 2006). For beach-spawning fishes, these ecosystems are critical for survival at the earliest stages of life. But beaches are also playgrounds and important recreation area for humans, and often are managed more for human safety and health than for their wilderness value (Bird, 1996; James, 2000). Beaches also have cultural legacies and the right of everyone to coastal access is protected and defended. However, with increasing human populations, pressures on all kinds of shorelines are growing as well (Defeo et al., 2009).

Fishes that spawn on beaches can be found on coastlines worldwide; on shores that are remote and shores that are highly developed; on beaches of rock, sand, gravel, or mud. Many of these fishes are endemic species, found only in very limited locations. Many require substrates that are scarce or patchy, or available only at certain tidal heights. Some require healthy vegetation of a particular type, that itself might be endangered. Each species has very specific requirements, or environmental specificity, for its critical breeding sites. These requirements include type of substrate, slope of the beach, wave energy, width of the beach, and tidal reach. Nests are often cryptic or hidden, yet they may be affected by disruption, egg predation, coastal construction, oil spills, and other human activities on the spawning grounds.

This chapter examines the larger context for conservation of critical coastal habitats over the long term. Some of these anthropogenic and natural factors threaten the future survival of beach-spawning fishes and the beaches themselves. One could write another chapter on the potential threats in the waters of the nearshore environment, an area that is home to many different kinds of marine organisms.

9.1 COASTAL CONSTRUCTION AND SHORELINE ARMORING ALTER BEACHES

Spawning sites of beach-spawning fishes are threatened by habitat loss from rising sea levels and encroaching coastal development. Construction of seawalls and buildings on the shore alter sediment transport and prevent natural replenishment of beaches after erosion (Orme et al., 2011). Beachfront construction changes the ability of the beach to respond to natural seasonal variations in erosion and deposition. Putting a structure on the back of the beach fixes it in one place and prevents natural retreat of the shoreline. Coupled with sea level rise and climate change, this leads to "coastal squeeze" (Figure 9.1). Coastal squeeze causes loss of habitat on beaches and alteration of wave regimes that beach-spawning fishes may require for oviposition and spawning behavior.

Shoreline modifications affect beach-spawning grounds for fishes (Robbins, 2006). Each will be considered separately, although often several of these may impact the same stretch of shoreline at the same time.

Construction of houses, port facilities, and other structures was an important factor in reduction of upper beach habitat historically (Griggs et al., 2005). Placing structures at fixed locations on a beach may prevent natural cycles of erosion and accretion of sand or gravel sediments and may necessitate additional construction to protect the existing structures later. The upper beach and coastal strand are the first habitat zones to be lost under these conditions, as waves lap up against the fixed structures even during moderate tides (National Research Council, 1995; Galofré et al., 2006).

Coastal construction often includes shoreline armoring. The installation of seawalls, revetments, rip rap, marinas, ports, and related structures has covered shoreline habitat and altered coastal processes along 30 percent of the southern California coast (Griggs et al., 2005; California Marine Life Protection Act Initiative, 2009). Approximately 30 percent of the shoreline in Puget Sound is developed or armored with boat ramps, bulkheads, buildings, seawalls, and other structures (Dugan & Hubbard, 2010). This alters wave regime, increases erosional loss of sand, and contributes to habitat loss in the intertidal zone (Dugan et al., 2008; Dugan & Hubbard, 2010).

FIGURE 9.1   Beaches are narrowing and losing habitat from sea level rise on one side and coastal construction on the other, creating "coastal squeeze." (Photo copyright 2002–2014, Kenneth and Gabrielle Adelman, California Coastal Records Project, http://www.californiacoastline.org. With permission.)

Construction on the shoreline has been more carefully regulated by legislation such as the Coastal Act in California and the Hydraulic Code in Washington in recent times that designate areas of special concern and require assessment of biological impacts before coastal areas are altered. Inadvertent impacts of shoreline development include bearing over the beach, altering the temperatures of the sand, and interrupting fine sediment erosion. Shorelines in Puget Sound that have less shade have lower survival of surf smelt eggs (Rice, 2006). However shorelines with excessive shade may slow development of California grunion embryos so that they are not ready to hatch at the next tidal cycle (Smyder & Martin, 2002).

Sand supply to beaches has dwindled over time. Seasonal erosion and deposition change beaches dynamically throughout the year in width and depth (Yates et al., 2009). Even so, in the past century, southern California has experienced a 50 percent reduction in the supply of sand to its beaches, because of the construction of dams and sand mining along rivers and on beaches. Beaches have narrowed and habitat has been lost as a result (Griggs et al., 2005; Orme et al., 2011). Efforts to remove dams and seawalls may increase sand supply locally in the future and allow natural replenishment of the beaches.

9.2   POLLUTION OF WATER AND LAND CAN HARM FISHES AT ALL LIFE STAGES

The early life stages of fishes are vulnerable to many different types of aquatic pollutants. As beach-spawning fishes tend to live near shore, they may come into contact with anthropogenic chemicals through industrial processes, accidental release, runoff from land, or disposal in sewage or other wastes that end up in the ocean. Endocrine disrupting chemicals may alter sexual development and reproduction in coastal fishes (Blewett et al., 2013).

The beach-spawning California grunion *Leuresthes tenuis* has been studied in bioassays for toxins including chlorpyrifos, carbophenothion, fenvalerate, thiobencarb, petroleum hydrocarbon, No. 2 fuel, benzo[a]pyrene, and tributylin (Winkler et al., 1983; Hose & Puffer, 1984; Borthwick et al., 1985; Clark et al., 1985; Goodman et al., 1985; Newton et al., 1985; McCoy, 1998). A classic study of developmental asymmetry was done on California grunion with p'-DDT (Valentine & Soule, 1973). A species is chosen for use in these types of assays because it shows a measurable response at lower levels than other species, the "canary in the coal mine" principle (Middaugh et al., 1993). All of these chemicals are present in urban runoff to the nearshore water and onto beaches where grunion spawn.

Chlorination may be used as water treatment for discharged water after sewage spills on or near beaches. This temporary action can be harmful to California grunion embryos and larvae in nearshore waters (Rosales-Casian et al., 1990).

Pollution from oil spills or fuel spills harms adult, larval, and embryonic grunion on shore and in the water (Gellert et al., 1994). More grunion than seabirds were killed by an oil spill in Southern California (American Trader Trustee Council, 2001). Herring eggs and larvae are vulnerable to danger of oil spills. Pacific herring *Clupea pallasii pallasii* spawn on surfgrass and other complex biogenic habitats near shore (see Chapter 4 for details). In 1989 in Alaska the spill from the oil tanker

Exxon Valdez was devastating to the local herring and the subsistence harvest of the herring spawn by indigenous Alaskans. In 2007 a spill of fuel oil from the cargo vessel *Cosco Busan* in the San Francisco Bay had strong negative impacts on herring survival and reproduction (Incardona et al., 2011), and its occurrence coincided with the local extirpation of an endemic population of California grunion (Martin et al., 2013).

We have observed that plastic trash that washes in on a beach can smother nest sites and kill embryos. Cigarette butts discarded in the sand accelerate the heart rate and increase the twitching activity of incubating embryos of *L. tenuis* (unpublished data).

9.3   SOME BEACH-SPAWNING FISHES ARE CAUGHT BY COMMERCIAL OR RECREATIONAL FISHING

Many beach-spawning fishes are fished at the time that they are most vulnerable, during their spawning runs, as discussed in Chapter 6. The practice of targeting spawning runs is shunned for most fish stocks because of its potentially devastating effects on a population (Tang & Chen, 2004). Taking a spawning adult not only removes that individual from the population, but also removes its potential offspring if taken before the spawning has occurred. When human populations were smaller, the numbers taken by a few individual anglers impacted the populations less. But today, the numbers of anglers can be quite large, and the near approach of these fishes to shore in spawning aggregations makes them an appealing target for anglers. No special training or equipment and no boats are needed to pick up the distracted spawning fishes. Regulations for recreational fishing of other species, such as closures, gear restrictions, and other limits, may not exist or may be enforced sporadically on beaches in remote locations.

Surf smelt are caught recreationally by hand and dip net during their spawning runs. In Puget Sound, they spawn year round. Only two miles of coastline are fished, but there are no restrictions on where they may be harvested. Even so several tons of surf smelt are caught each year. In the past, this species was so abundant that a hundred anglers could fish on the same beach, but now they are much less abundant and catches are far below historic levels for all forage fishes in Puget Sound. Traditional stock management techniques do not work well for forage fishes (Bargmann, 1998; Benoît & Swain, 2008; Pikitch et al., 2014). The majority of young surf smelt may be produced on only a few beaches (Quinn et al., 2012).

Following the collapse of the cod fishery in the late twentieth century in the North Atlantic, populations of their prey species, including herring and sand lance, increased (Frank et al., 2011, 2013). These are different species than those found in the Pacific Ocean but they have a similar trophic level in the ecosystem. As the cod population has slowly recovered, the populations of forage fishes have started to decrease.

Commercial fishing tends to focus on larger species that are predatory on smaller fishes. A commonly voiced concern as human population grows and fishing methods become more efficient is that the top predators are being depleted from the ocean, and the marine food webs are changing as a result (Pauly et al., 1998; Estes et al., 2011; Pauly & Palomares, 2005). With fewer large fish available, commercial fishing may target the more abundant fishes that were food for that top predator.

While this keeps the total size of the catch high, the individual fishes caught are smaller. Additional fishing pressure on the forage fishes may in turn make it more difficult for the overexploited fishes to recover as they are now competing with the fishing vessels for their food. At the same time, the forage fishes are literally caught in the middle, taken by both sides.

The Atlantic capelin *Mallotus villosus* population collapsed at about the same time as the cod in the early 1990s (Obradovich et al., 2013), declining to their lowest level in 500 years. Capelin are the key prey item of cod, as well as a commercial fishery, and rebuilding the cod population depends on rebuilding the capelin population. However, this species has not bounced back with the loss of the cod (Carscadden et al., 2013).

The situation is complex because forage fishes and young cod occupy different levels of the marine food web at different stages of life. In addition multiple species may be in competition for the same prey resources, or may alternate between being predator and prey at different life cycle stages. In the Barents Sea, Atlantic herring, cod, and capelin form an interlinked food web, with the biomass of each species fluctuating tremendously across time (Figure 9.2). The capelin stock collapsed in the

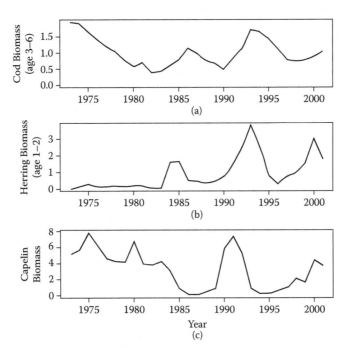

FIGURE 9.2 The biomass in 10^6 tons of (a) adult cod *Gadus morhua*, age 3–6 years; (b) Atlantic herring *Clupea harengus*, age 1–2 years, that prey on capelin larvae; and (c) capelin *Mallotus villosus* in the Barents Sea north of Scandinavia and Russia. Capelin stock has collapsed in each of the last three decades, resulting in fishing moratoriums. (From Hjermann, D. O., Ottersen, G., & Stenseth, N. C., *Proceedings of the National Academy of Sciences*, 101, 2004 11679–11684, http://www.pnas.org/cgi/doi/10.1073/pnas.0402904101. Copyright 2004, National Academy of Sciences, United States. With permission.)

mid-1980s and mid-1990s, and again in 2004, with loss of 85–95 percent compared with previous years (Hjermann et al., 2004). Fishing moratoriums are complicated by the effects of predation by herring and cod on the capelin. Management of any one stock depends on the careful analysis of the other two species, and a multispecies or community-level approach is needed for sustainability of these important fisheries (Hjermann et al., 2004), and their seabirds and marine mammal predators (Kaschner et al., 2006).

California grunion and the remains of other beach-spawning species have been found in middens from prehistoric cultures in California (Gobalet & Jones, 1995). These fishes completely emerge from water onto sandy beaches during their spawning runs, and catching them requires no boat and no special technical skill; in fact, no gear is legal for catching the fish; only the bare hands can be used (Spratt, 1986). Literally thousands of hopeful anglers of all ages line the shores of some southern California beaches on nights of predicted runs, hoping to capture some fish. The Gulf grunion in Baja California are under similar pressure and even more vulnerable in the daylight (Figure 9.3). Changes in seasonal temperatures or oceanographic conditions can alter spawning seasons and leave fixed closure times that bear little relevance for protection of spawning runs (Martin et al., 2013).

The fishery for whitebait, *Galaxias maculatus,* in New Zealand targets the migrating juveniles rather than the spawning adults (McDowall, 1996). Although the catch fluctuates wildly, establishing regulations on this artisanal fishery in rural conditions is extremely challenging (McDowall, 2006), even though the species is declining. See Chapter 10 for more on conservation efforts for this fishery and species.

FIGURE 9.3 (See color insert.)   Recreational anglers on shore take fish during a spawning run of Gulf grunion. (Photo by Tim Bourque. With permission.)

9.4 MAINTENANCE ACTIVITIES AND MANAGEMENT ACTIONS CAN AFFECT SPAWNING SITES ON RECREATIONAL BEACHES

Beach recreation and tourism generate literally billions of dollars in revenues (Pendleton et al., 2001; Klein et al., 2004). The California grunion spawn on these beaches at the same time of year that millions of visitors come out to play, during late spring and summer. Although the spawning runs occur late at night and have some regulatory protection in the form of closed season and gear restrictions, until recently little attention was given to protection of the embryos left behind.

While buried under just a few inches of sand, grunion nests are not visible from the surface. They are vulnerable to vehicular traffic, beach grooming and raking (Figure 9.4), digging of sand, and burial from sand transport (Martin et al., 2006). They can also be washed out by hard rain or high wave action during storms, runoff from creeks and streams that breach berms, or if a creek mouth widens or migrates or is artificially breached during the incubation period.

Mechanical beach grooming removes kelp wrack and disturbs natural beach habitat on 45 percent of Southern California beaches (California Marine Life Protection Act Initiative, 2009). If raking and grooming occur in the habitat zone of the grunion nests, eggs are destroyed (Martin et al., 2006). Invertebrates are also injured by beach grooming (Llewellyn & Shackley, 1996; Dugan et al., 2000). This degrades the habitat function for shorebirds and birds migrating along the Pacific Flyway (Dugan & Hubbard, 2009; Dugan et al., 2003, 2011). The constant turnover of the upper part of the substrate prevents plants from rooting and limits the development of coastal strand habitat (Dugan & Hubbard, 2009).

In winter, some municipalities order the construction of artificial berms by bulk movement of sand from the lower beach to the upper beach. This action is usually

FIGURE 9.4 Beach grooming with mechanized maintenance in San Diego. (Photo by Karen Martin.)

done to protect buildings from surging waves during winter storms. This causes disturbance to the intertidal biota at the time the berms are constructed, by removing habitat and removing or smothering infauna. Later in spring, when the berms are flattened and pushed back down the face of the beach, additional disruption occurs when the biota on the lower beach is buried under a thick layer of sand. The timing of the building and pushing down of the berms can affect many organisms, including grunion eggs, hidden under the surface of the sand.

Beach sand is unconsolidated sediment that moves and undergoes cycles of erosion and accretion (McLachlan & Brown, 2006). Over time beaches erode, both by natural and anthropogenic factors (Orme et al., 2011). One common management action is to obtain sand from another source and spread it over the area. This sand replenishment is sometimes called beach nourishment, beach fill, or beach restoration, but unlike other forms of habitat restoration, this usually does not entail any subsequent biological replacement or habitat monitoring activities (Defeo et al., 2009; Dugan et al., 2010). Sand replenishment may temporarily increase the width of beaches and provide more habitat area, but the new sediments typically do not remain for more than a few years to a decade.

Typical replenishment practices result in large disturbance to fauna (Peterson & Bishop, 2005; Peterson et al., 2006). Recovery times may be on the order of a few years for invertebrate species with planktonic larvae if the sediment size is appropriate. Species with low dispersal abilities may take considerably longer (Peterson et al., 2000). If nourishment is repeated at short intervals, such as every few years, full recovery of the ecosystem may not be achieved (Schlacher et al., 2008).

Beach replenishment involves depositing wet sand on a beach either by the truckload or with a pipeline from an ocean dredge source, followed by grading the sand with heavy equipment. This construction destroys grunion eggs if they are present, and may also interfere with the adults during a spawning run (personal observation). The pumping equipment and the shoreline construction create disturbance and hazards by altering wave energy and creating traps behind sand or barriers from which the spawning fish cannot escape. Grunion eggs that are produced may be buried deeply under too much sand, either by direct grading or by increased sand transport by waves as the new sand settles, and may not be able to hatch or survive (Lawrenz-Miller, 1991). Eventually, California grunion will return to spawn on replenished or manmade beaches (Clark, 1928; Green, 2002; Martin et al., 2013), while they last.

Vehicles on beaches can crush or trample the biota during construction and sand replenishment. Off-road vehicle use including lifeguards, vendors, public safety, and recreational vehicles can trample and crush wrack and the associated biota. We have occasionally observed public safety officers literally driving across an active grunion run, killing adult fish. In the Gulf of California, there have been reports of people driving quad-runners over the daytime spawning runs of Gulf grunion (T. Bourke, personal communication). When vehicles drive over nest sites, buried grunion eggs can be killed by crushing or by being turned up to the surface and drying out in the sun.

9.5   INDUSTRIAL ACTIVITIES NEAR BEACHES CAN IMPACT FISHES

Another area of potential threats to the beaches that are critical for these spawning fishes is development for industrial uses. Historically the coasts have been modified and used for many types of industrial practices not only including marinas and harbors, but also for manufacturing or processing in factories and power plants. Current efforts to increase alternative energy sources and aquaculture facilities may lead to additional pressures on shorelines in the future.

Power plant effluent reduces survival of larval California grunion *L. tenuis* (Ehrlich, 1977). Many larvae of this and other species are lost by impingement and entrapment when power plants use powerful suction to take in seawater to cool their equipment (Miller & Schiff, 2012). The same kind of entrapment may occur at desalination plants along the coast that draw in ocean water for conversion to drinking water.

Grunion larvae are sensitive to changes in salinity (Reynolds et al., 1976). Desalination plants are being planned on the California coast to make drinking water from seawater, and are already in use in many locations on a relatively small scale. These produce hypersaline brine at least twice the concentration of seawater (Service, 2006), which is released back into the ocean (Hashim & Hajjaj, 2006). If discharged near shore, it could be harmful or fatal to grunion embryos and larvae (Matsumoto & Martin, 2008).

Efforts to develop wave energy have not yet been successful on an industrial scale. However, the templates that are being tested could potentially impact spawning migrations to beaches. A large array of equipment in an area with consistent wave activity has the potential to interfere with movement between shore and deeper water for fishes and other animals that traverse different ecosystems during the course of a day or season (Boehlert et al., 2007), including anadromous fishes such as salmon.

In addition the intent of capturing some of the energy from the wave in order to transform it into electricity for storage and later use will also, by the laws of physics, alter the energy of the wave. Many beach-spawning fishes rely on tidal cues and waves to synchronize their spawning activities, and disruption of these signals could reduce spawning success in the areas with these installations. The issues of impingement and entrapment have not been addressed yet for these structures but these are also concerns.

Another concern for coastal ecosystems is the growth of aquaculture facilities. In Puget Sound, some areas are in jeopardy because the aquaculture industry is in the process of developing shellfish farms (Dan Penttila, personal communication). Other marine life is removed to make habitats suitable for aquaculture facilities, including eelgrass and ghost shrimp in areas with finer sediments.

9.6   CLIMATE CHANGE AND SEA LEVEL RISE MAY CAUSE LOSS OF BEACH HABITAT

Global warming can contribute to loss of coastal habitat through sea level rise and by increased extreme weather events such as severe storms and hurricanes (Zhang et al., 2004). Water expands when heated, and this, coupled with melted sea ice, is predicted to cause increased water levels along shorelines worldwide in the next decade (Cayan et al., 2008).

Loss of shore habitat is a likely consequence of sea level rise, especially when coupled with coastal erosion (Feagin et al., 2005; Schlacher et al., 2008). Shifting sands and seasonal changes in beach width are part of the normal response of the coast to waves and weather (Flick, 1993; Griggs et al., 2005; Yates et al., 2009). However, there is concern that global shifts in climate, even if gradual, may trigger increased numbers of severe weather events that can lead to dramatic loss of sediments on beaches and prolonged, difficult periods of recovery that may result in local extinctions of sensitive species (Harley & Paine, 2010).

Geomorphologists have begun to calculate the need for sand replenishment to restore beach width in hopes of retaining some recreational beaches after the consequences of sea level rise in the coming years (Flick & Ewing, 2009). However, the need is projected to be so great that sources of sand are increasingly difficult to identify, meaning that the projected costs will likely rise (Smith et al., 2009).

All beaches are geologically young and relatively unstable (McLachlan & Brown, 2006). Under natural ecological succession, as the sea level changes, the shoreline retreats or extends from its previous location as sediments undergo cycles of deposition and erosion. The organisms on the beach survive by shifting as the habitat slowly changes. An unaltered coastline is able to respond dynamically to changes in sea level rise (Revell et al., 2011), but now a large portion of the coast of many areas is urbanized, armored, and developed (Liedersdorf et al., 1994; Orme et al., 2011). The percent area with good viability and ecological integrity currently is a small fraction of the total shoreline.

Some parks that include beaches or other undeveloped areas have potential for managed retreat of the shoreline by working within natural processes of shoreline reshaping. This does not preserve structures, as seawalls and coastal armoring are intended to do, but it would preserve habitat. Especially on coastlines where beaches are patchy and widely separated by intervening coastal bluffs or armoring, recognizing the importance of these areas with somewhat wild coasts is vital to the survival of the animals and plants dependent on the beaches (James, 2000; Defeo et al., 2009; Dugan et al., 2010).

One possible response that fishes may have to the changes in beaches with sea level rise would be to find other locations for spawning. If temperatures or surf waves become too extreme, if the higher zones in the intertidal are lost to erosion and shoreline armoring, if the beach substrates change through loss by erosion and lack of replacement, then perhaps these fishes that have already demonstrated their behavioral plasticity will just take the party somewhere else.

We know already that some beach-spawning species also spawn in marine habitats underwater. For example, the midshipman *Porichthys notatus* spawns intertidally and to 80 m depth, but not much is known about the differential in embryo survival in each habitat (Feder et al., 1974). Capelin in Newfoundland spawn on a beach drowned at end of last glaciation, but few offspring survive the underwater incubation as opposed to the many that survive in intertidal zone (Nakashima & Taggart, 2002). This species has a circumpolar distribution, but at least since the mid twentieth century capelin disappeared from central and southern British Columbia

(Dan Penttila, personal communication). Therefore, it is not clear that spawning habitat shifts are easy, or even possible, for fishes within a geographic area (Krueger et al., 2010). In addition, along with climate change the predator populations will be shifting (Hazen et al., 2013), and new combinations within marine food webs add greater uncertainty to the move.

Climate change may alter plankton abundance due to temperature shifts that may affect the recovery of *Mallotus villosus* (Obradovich et al., 2013). Changing temperatures may affect the survival of incubating embryos on beaches and elsewhere (Nakashima & Wheeler, 2002; Nakashima & Taggart, 2002) sooner, and in ways that differ from the effects on the adults at sea. Sympatric species that depend on the same food resources, such as forage fishes, may have different optimum temperatures at different life cycle stages that confer advantages or disadvantages in this competition (Takasuka et al., 2007).

The California grunion provide a cautionary tale. This endemic species spawns only on beaches and is found solely in California and Baja California. More than 95 percent of the species resides on the coast of Southern California, with a few very small populations found occasionally in bays north of Pt. Conception (Roberts et al., 2007; Johnson et al., 2009; Sandrozinsky, 2013). Their Southern California beach homes are on a coastline that is heavily impacted by human development and recreation (Griggs et al., 2005; Orme et al., 2011).

Because of their complete reliance on beach spawning, California grunion (along with their congeners, the Gulf grunion) are exceptionally vulnerable to habitat loss along coastlines. They are not known to spawn anywhere else but beaches. The nests are restricted to a narrow band between the mean high tide line and the highest high tide line for their critical reproductive habitat. This zone is usually less than 3 m wide, sometimes only a few inches wide, and it shifts as the beach face changes over time.

A rough, simple accounting of available habitat for this species can be calculated according to information provided by the California Marine Life Protection Act Regional Profile of the South Coast District of California (2009). In this region, from Pt. Conception to the Mexican border, the sandy beach comprises approximately 380 linear miles of shoreline. The width of the grunion nesting zone is at most 3 m, the correct tidal height on shore on a very flat beach. Often the run is much narrower. With the best case scenario, the total area available along the entire South Coast District is 3 m or 0.00186 miles wide, on a discontinuous linear shoreline of 380 miles. That works out to the equivalent of only 0.71 square miles. This tiny area, smaller than the size of Central Park in New York City, is the entire space available for nesting for more than 95 percent of the population of this endemic species. All of it is influenced by events on land and sea, and this small amount is present as even smaller, discontinuous patches along a coast that has many different management jurisdictions and human activities. Loss of any of this already miniscule critical habitat is cause for alarm.

Habitat expansion to the north is not a simple solution. Between 2001 and 2008, California grunion expanded their habitat northward and established disjunct populations hundreds of kilometers north of Pt. Conception, in Monterey Bay, San Francisco Bay, and Tomales Bay (Roberts et al., 2007; Johnson et al., 2009; Byrne et al., 2013). Even though these populations were able to grow and breed, the individual fishes were significantly smaller at maturity, produced significantly fewer eggs in their clutches, and lived shorter lives (Martin et al., 2013). All three populations disappeared before the end of the decade. These areas may be recolonized in the future, but the difficulties the species faces will make habitat expansion more complex and uncertain than simply a shift poleward.

9.7   SUMMARY OF NEW THREATS TO BEACH-SPAWNING FISHES

The anthropogenic impacts on beaches are beginning to be studied, and they are extensive. Continuing exposure to these threats could affect most of the breeding areas and critical habitats for beach-spawning fishes. Coastal development alters wave regimes and sedimentation. Changes to vegetation may result in degradation or loss of spawning habitat, or changes in shade cover and shore temperatures. Pollution impacts reproductive success, and climate change threatens the delicate balance of humidity and temperature that the developing eggs require on beaches. Sea level rise coupled with shoreline armoring creates a "coastal squeeze" of habitat loss from both directions for many narrow beaches that no longer have adjacent upland area for subsidence or retreat.

Subject to both terrestrial and marine influences, the beach is truly the transition on the edge between both. With seasonal deposition and erosion, this habitat is ever changing. Use of a particular stretch of beach by spawning fishes may be sporadic over a season or between years, depending on the conditions. Very few areas are protected on behalf of these fishes, and management regulations are not enforced consistently. Most of these species do not survive long or reproduce in captivity and must complete their life cycles in the wild. If lost, these species cannot be rescued, as there are no captive populations to draw from and no other places to find that endemic species if its home disappears. Chapter 10 examines some conservation efforts to preserve these beach-spawning fishes around the globe.

REFERENCES

American Trader Trustee Council. (2001). Final restoration plan and environmental assessment for seabirds injured by the American Trader oil spill. *Report of the American Trader Natural Resource Trustee Council*, U.S. Fish and Wildlife Service, California Department of Fish and Game, and National Oceanic and Atmospheric Administration.

Bargmann, G. (1998). *Forage Fish Management Plan*. Olympia, WA: Washington Department of Fish and Wildlife, 67 pp.

Benoît, H. P. & Swain, D. P. (2008). Impacts of environmental change and direct and indirect harvesting effects on the dynamics of a marine fish community. *Canandian Journal of Fisheries and Aquatic Sciences* 65, 2088–2104.

Bird, E. (1996). *Beach Management*. Chichester, England: Wiley & Sons, 281 pp.

Blewett, T. A., Robertson, L. M., Machatchy, D. L. & Wood, C. M. (2013). Impact of environmental oxygen, exercise, salinity, and metabolic rate on the uptake and tissue-specific distribution of 17d-ethyngl-estradiol in the euryhaline teleost *fundulus heteroclitus*. *Aquatic Toxicology* 138, 43–51.

Boehlert, G. W., McMurray, G. R. & Tortorici, C. E. (2007). *Ecological Effects of Wave Energy Development in the Pacific Northwest*. NOAA Technical Memorandum NMFS-F/SPO-92, 174 pp.

Borthwick, P. W., Patrick, J. M. Jr. & Middaugh, D. P. (1985). Comparative acute sensitivities of early life stages of Atherinid fishes to chlorpyrifos and thiobencarb. *Archives of Environmental Contamination and Toxicology* 14, 465–473.

Byrne, R., Bernardi, G. & Avise, J. C. (2013). Spatiotemporal genetic structure in a protected marine fish, the California grunion (*Leuresthes tenuis*), and relatedness in the genus *Leuresthes*. *Journal of Heredity* 104, 521–531. DOI: 10.1093/jhered/est024.

California Marine Life Protection Act Initiative (2009). *Regional Profile of the MLPA South Coast Study Region (Point Conception to the California–Mexico Border)*. California Natural Resources Agency, Sacramento, CA. 196 pp. plus maps and appendices. http://www.dfg.ca.gov/marine/mpa/scprofile.asp.

Carscadden, J. E., Gjøsæter, H., & Vilhjálmsson, H. (2013). A comparison of recent changes in distribution of capelin (*Mallotus villosus*) in the Barents Sea, around Iceland and in the Northwest Atlantic. *Progress in Oceanography* 114, 64–83.

Cayan, D. M., Bromirski, P. D., Hayhoe, K., Tyree, M., Dettinger, M. D. & Flick, R. E. (2008). Climate change projections of sea level extremes along the California coast. *Climatic Change* 87, Suppl. 1, S57–S73.

Clark, F. N. (1928). Grunion on Cabrillo Beach. *California Fish & Game* 14, 4, 273–274.

Clark, J. R., Patrick, J. M. Jr., Middaugh, D. P. & Moore, J. C. (1985). Relative sensitivity of six estuarine fishes to carbophenothion, chlorpyrifos, and fenvalerate. *Ecotoxicology and Environmental Safety* 10, 382–390.

Defeo, O., McLachlan, A., Schoeman, D. S., Schlacher, T. A., Dugan, J., Jones, A., Lastra, M. & Scapini, F. (2009). Threats to sandy beach ecosystems: a review. *Estuarine, Coastal and Shelf Science* 81, 1–12.

Dugan, J. E. & Hubbard, D. M. (2009). Loss of coastal strand habitat in Southern California: the role of beach grooming. *Estuaries and Coasts* 33, 67–77. DOI: 10.1007/s12237-009-9239-8.

Dugan, J. E. & Hubbard, D. M. (2010). Ecological effects of coastal armoring: a summary of recent results for exposed sandy beaches in Southern California. In *Puget Sound Shorelines and the Impacts of Armoring: Proceedings of a State of the Science Workshop*, 187–193. USGS Publications.

Dugan, J. E., Defeo, O., Jaramillo, E., Jones, A. R., Lastra, M., Nel, R., Peterson, C. H., Scapini, F., Schlacher, T. & Schoeman, D. S. (2010). Give beach ecosystems their day in the sun. *Science* 329, 1146.

Dugan, J. E., Hubbard, D. M., McCrary, M. D. & Pierson, M. O. (2003). The response of macrofauna communities and shorebirds to macrophyte wrack subsidies on exposed sandy beaches of Southern California. *Estuarine, Coastal and Shelf Sciences* 58S, 25–40.

Dugan, J. E., Hubbard, D. M., Rodil, I. F., Revell, D. L. & Schroeter, S. (2008). Ecological effects of coastal armoring on sandy beaches. *Marine Ecology* 29, S1, 160–170. DOI: 10.1111/j.1439-0485.2008.00231.x.

Dugan, J., Hubbard, D., Page, H. & Schimel, J. (2011). Marine macrophyte wrack inputs and dissolved nutrients in beach sands. *Estuaries and Coasts* 34, 839–850.

Dugan, J. E., Hubbard, D. M., Engle, J. M., Martin, D. L., Richards, D. M., Davis, G. E., Lafferty, K. D. & Ambrose, R. F. (2000). Macrofauna communities of exposed sandy beaches on the Southern California mainland and Channel Islands. In *Fifth California Islands Symposium*, 339–346. Outer Continental Shelf Study, US Department of the Interior, Minerals Management Service 99-0038. http://science.nature.nps.gov/im/units/medn/symposia/5th%20California%20Islands%20Symposium%20%281999%29/Proceedings/.

Ehrlich, K. F. (1977). Inhibited hatching success of marine fish eggs by power plant effluent. *Marine Pollution Bulletin* 8, 228–229.

Estes, J. A., Terborgh, J., Brashares, J. S., Power, M. E., Berger, J., Bond, W. J., Carpenter, S. R., et al. (2011). Trophic downgrading of planet Earth. *Science* 333, 301–306.

Feagin, R. A., Sherman, D. J. & Grant, W. E. (2005). Coastal erosion, global sea-level rise, and the loss of sand dune plant habitats. *Frontiers in Ecology and the Environment* 7, 3, 359–364.

Feder, H. M., Turner, C. H. & Limbaugh, C. (1974). *Observations on Fishes Associated with Kelp Beds in Southern California. Fish Bulletin 160.* State of California Resources Agency, Department of Fish & Game. http://content.cdlib.org/view?docId=kt9t1nb3sh&brand=oac4.

Flick, R. E. (1993). The myth and reality of Southern California beaches. *Shore & Beach* 61, 3–13.

Flick, R. E. & Ewing, L. C. (2009). Sand volume needs of Southern California beaches as a function of future sea-level rise rates. *Shore & Beach* 77, 36–45.

Frank, K. T., Leggett, W. C., Petrie, B. D., Fisher, J. A. D., Shackell, N. L. & Taggart, C. T. (2013). Irruptive prey dynamics following the groundfish collapse in the Northwest Atlantic: an illusion? *ICES Journal of Marine Science* 70, 1299–1307. DOI: 10.1093/icesjms/fst111.

Frank, K. T., Petrie, B., Fisher, J. A. D. & Leggett, W. C. (2011). Transient dynamics of an altered large marine ecosystem. *Nature* 477, 86–89.

Galofré, J., Medina, R., González, M. & Montoya, F. J. (2006). Influence of a marina construction on adjacent beaches: "Roda de Bara" case study. *Coastal Engineering* 2006, 3228–3240.

Gellert, G. A., Daugherty, S. J., Rabiee, L., Mazur, M. & Merryman, R. E. (1994). California's American Trader oil spill: effective interagency and public–private collaboration in environmental disaster. *Journal of Environmental Health* 46, 4, 7–additional.

Gobalet, K. W. & Jones, T. L. (1995). Prehistoric Native American fisheries of the central California coast. *Transactions of the American Fisheries Society* 124, 813–823.

Goodman, L. R., Hansen, D. J., Cripe, G. M., Middaugh, D. P. & Moore, J. C. (1985). New early life-stage toxicity test using the California grunion (*Leuresthes tenuis*) and results with chlorpyrifos. *Ecotoxicology and Environmental Safety* 10, 12–21.

Green, K. (2002). *Beach Nourishment: A Review of the Biological and Physical Impacts.* Atlantic States Marine Fisheries Commission Habitat Management Series #7. http://www.asmfc.org/publications/habitat/beachNourishment.pdf.

Griggs, G., Patsch, K. & Savoy, L. (2005). *Living with the Changing California Coast.* Berkeley: University of California Press, 551 pp.

Harley, C. D. G. & Paine, R. T. (2010). Contingencies and compounded rare perturbations dictate sudden distributional shifts during periods of gradual climate change. *Proceedings of the National Academy of Sciences* 106, 11172–11176.

Hashim, A. & Hajjaj, M. (2005). Impact of desalination plants fluid effluents on the integrity of seawater with the Arabian Gulf in perspective. *Desalination* 182, 373–383.

Hazen, E. L., Jorgensen, S., Rykaczewski, R. R., Bograd, S. J., Foley, D. G., Jonsen, I. D., Shaffer, S. A., Dunne, J. P., Costa, D. P., Crowder, L. B. & Block, B. A. (2013). Predicted habitat shifts of Pacific top predators in a changing climate. *Nature Climate Change* 3, 234–238. DOI: 10.1038/nclimate1686.

Hjermann, D. O., Ottersen, G. & Stenseth, N. C. (2004). Competition among fishermen and fish causes the collapse of the Barents Sea capelin. *Proceedings of the National Academy of Sciences* 101, 11679–11684. www.pnas.org/cgi/doi/10.1073/pnas.0402904101.

Hose, J. E. & Puffer, H. W. (1984). Oxygen consumption rates of grunion (*Leuresthes tenuis*) embryos exposed to the petroleum hydrocarbon, benzo[a]pyrene. *Environmental Research* 35, 413–420.

Incardona, J. P., Vines, C. A., Anulacion, B. F., Baldwin, D. H., Day, H. L., French, B. L., Labenia, J. S., et al. (2011). Unexpectedly high mortality in Pacific herring embryos exposed to the 2007 *Cosco Busan* oil spill in San Francisco Bay. *Proceedings of the National Academy of Sciences Plus* 2011. DOI: /10.1073/pnas.1108884109.

James, J. J. R. (2000). From beaches to beach environments: linking the ecology, human use, and management of beaches in Australia. *Ocean and Coastal Management* 43, 495–514.

Johnson, P. B., Martin, K. L., Vandergon, T. L., Honeycutt, R. L., Burton, R. S. & Fry, A. (2009). Microsatellite and mitochondrial genetic comparisons between northern and southern populations of California grunion *Leuresthes tenuis*. *Copeia* 2009, 467–476.

Kaschner, K., Karpouzi, Y., Watson, R. & Pauly, D. (2006). Forage fish consumption by marine mammals and seabirds. In *On the Multiple Uses of Forage Fish: From Ecosystems to Markets* (Alder, J. & Pauly, D., eds.), 33–46. *Fisheries Centre Research Reports* 14(3). British Columbia, Canada: Fisheries Centre, University of British Columbia.

Klein, Y. L., Osleeb, J. P. & Viola, M. R. (2004). Tourism-generated earnings in the coastal zone: a regional analysis. *Journal of Coastal Research* 20, 1080–1088.

Krueger, K. L., Pierce Jr., K. B., Quinn, T. & Penttila, D. (2010). Anticipated effects of sea level rise in Puget Sound on two beach-spawning fishes. In *Puget Sound Shorelines and the Impacts of Armoring: Proceedings of a State of the Science Workshop*, 266. U. S. Geological Survey.

Lawrenz-Miller, S. (1991). Grunion spawning versus beach nourishment: nursery or burial ground? *Proceedings of the 7th Symposium on Coastal and Ocean Management. Coastal Zone '91* 3, 2197–2208.

Leidersdorf, C., Holler, R. C. & Woodell, G. (1994). Human intervention with the beaches of Santa Monica Bay, California. *Shore & Beach* 62, 29–38.

Llewellyn, P. J. & Shackley, S. E. (1996). The effects of mechanical beach-cleaning on invertebrate populations. *British Wildlife* 7, 147–155.

Martin, K. L. M., Heib, K. A. & Roberts, D. A. (2013). A Southern California icon surfs north: local ecotype of California grunion, *Leuresthes tenuis* (Atherinopsidae) revealed by multiple approaches during temporary habitat expansion into San Francisco Bay. *Copeia* 2013, 729–739.

Martin, K., Speer-Blank, T., Pommerening, R., Flannery, J. & Carpenter, K. (2006). Does beach grooming harm grunion eggs? *Shore & Beach* 74, 17–22. DOI: 10.1643/CI-13-036.

Matsumoto, J. K. & Martin, K. L. M. (2008). Lethal and sublethal effects of altered sand salinity on embryos of beach-spawning California grunion. *Copeia* 2008, 483–490. DOI: 10.1643/CP-07-097.

McCoy, E. A. (1998). *Effects of No. 2 Fuel on* Leuresthes tenuis *and Comparison to P450 Reporter Gene System Biomarker*. MS Thesis, San Diego State University.

McDowall, R. M. (1996). Managing the New Zealand whitebait fishery: a critical review of the role and performance of the Department of Conservation. No. 32 NIWA Science and Technology Series. Wellington, NZ: National Institute of Water and Atmospheric Research Ltd. http://docs.niwa.co.nz/library/public/NIWAsts32.pdf.

McDowall, R. M. (2006). Crying wolf, crying foul, or crying shame: alien salmonids and a biodiversity crisis in the southern cool-temperate galaxoid fishes? *Reviews in Fish Biology and Fisheries* 16, 233–422. DOI: 10.1007/s11160-006-9017-7.

McLachlan, A. & Brown, A. C. (2006). *The Ecology of Sandy Shores, 2nd edition*. San Diego, CA: Academic Press.

Middaugh, D. P., Goodman, L. R. & Hemmer, M. J. (1993). Methods for spawning, culturing and conducting toxicity tests with early life stages of estuarine and marine fishes. In *Handbook of Ecotoxicology, Volume One* (Calow, P., ed.), 167–192. Oxford, UK: Blackwell Scientific Publishers.

Miller, E. F. & Schiff, K. (2012). Descriptive trends in Southern California Bight demersal fish assemblages since 1994. *CalCOFI Reports* 53, 107–131.

Nakashima, B. S. & Taggart, C. T. (2002). Is beach-spawning success for capelin, *Mallotus villosus* (Muller), a function of the beach? *ICES Journal of Marine Science* 59, 897–908.

Nakashima, B. S. & Wheeler, J. P. (2002). Capelin (*Mallotus villosus*) spawning behavior in Newfoundland waters: the interaction between beach and demersal spawning. *ICES Journal of Marine Science* 59, 909–916.

National Research Council. (1995). *Beach Nourishment and Protection*. Washington, DC: National Academy Press.

Newton, F., Thum, A., Davidson, B., Valkirs, A. & Seligman, P. (1985). Effects on the growth and survival of eggs and embryos of the California grunion (*Leuresthes tenuis*) exposed to trace levels of tributylin. *Naval Oceans Systems Center, Technical Report 1040*. San Diego, CA.

Obradovich, S. G., Carruthers, E. H. & Rose, G. A. (2013). Bottom-up limits to Newfoundland capelin (*Mallotus villosus*) rebuilding: the euphausiid hypothesis. *ICES Journal of Marine Science* 71, 775–783. DOI: 10.1093/icesjms/fst184.

Orme, A. R., Griggs, G. B., Revell, D. L., Zoulas, J. G., Grandy, C. C. & Koo, H. (2011). Beach changes along the Southern California coast during the 20th century: a comparison of natural and human forcing factors. *Shore & Beach* 79, 38–50.

Pauly, D. & Palomares, M. L. (2005). Fishing down marine food webs: it is far more pervasive than we thought. *Bulletin of Marine Science* 76(2), 197–211.

Pauly, D., Christensen, V., Dalsgaard, J., Froese, R. & Torres, F. (1998). Fishing down marine food webs. *Science* 279, 860–863.

Pendleton, L., Martin, N. & Webster, D. G. (2001). Public perceptions of environmental quality: a survey study of beach use and perceptions in Los Angeles County. *Marine Pollution Bulletin* 42, 1155–1160.

Peterson, C. H. & Bishop, M. J. (2005). Assessing the environmental impacts of beach nourishment. *Bioscience* 55, 10, 887–896.

Peterson, C. H., Bishop, M. J., Johnson, G. A., D'Anna, L. M. & Manning, L. M. (2006). Exploiting beach filling as an unaffordable experiment: benthic intertidal impacts propagating upwards to shorebirds. *Journal of Experimental Marine Biology and Ecology* 338, 206–221.

Peterson, C. H., Hickerson, D. H. M. & Johnson, G. G. (2000). Short-term consequences of nourishment and bulldozing on the dominant large invertebrates of a sandy beach. *Journal of Coastal Research* 16, 368–378.

Pikitch, E. K. et al. (2014). The global contribution of forage fishes to marine fisheries and ecosystems. *Fish and Fisheries* 15, 43–64. DOI: 10.1111/faf.12004.

Quinn, T., Krueger, K., Pierce, K., Penttila, D., Perry, K., Hicks, T., & Lowry, D. (2012). Patterns of surf smelt, *Hypomesus pretiosus*, intertidal spawning habitat use in Puget Sound, Washington State. *Estuaries and Coasts* 35, 1214–1228. DOI: 10.007/s12237-012-9511-1.

Revell, D. L., Dugan, J. E. & Hubbard, D. M. (2011). Physical and ecological responses to the 1997–98 El Nino. *Journal of Coastal Research* 27, 718–730.

Reynolds, W. W., Thomson, D. A. & Casterlin, M. E. (1976). Temperature and salinity tolerances of larval California grunion, *Leuresthes tenuis* (Ayres): a comparison with Gulf grunion, *L. sardina* (Jenkins & Evermann). *Journal of Experimental Marine Biology and Ecology* 24, 73–82.

Rice, C. (2006). Effects of shoreline modification on a northern Puget Sound beach: microclimate and embryo mortality in surf smelt (*Hypomesus pretiosus*). *Estuaries and Coasts* 29, 63–71.

Robbins, E. (2006). *Essential Fish Habitat in Santa Monica Bay, San Pedro Bay, and San Diego Bay: A Reference Guide for Managers*. MS Thesis, Duke University.

Roberts, D., Lea, R. N. & Martin, K. L. M. (2007). First record of the occurrence of the California grunion, *Leuresthes tenuis*, in Tomales Bay, California: a northern extension of the species. *California Fish & Game* 93, 107–110.

Rosales-Casian, J. A., Alfonso-Hernandez, I. & Hammann, N. G. (1990). The effect of seawater chlorination on the survival and growth of California grunion (*Leuresthes tenuis* Ayres) larvae, in laboratory conditions. *Ciencias Marinas* 16, 31–46.

Sandrozinsky, A. (2013). Status of the fishery: California grunion. *Status of the Fisheries Report*, California Department of Fish and Wildlife. http://www.dfg.ca.gov/marine/status.

Schlacher, T. A., Schoeman, D. S., Dugan, J., Lastra, M., Jones, A., Scapini, F. & McLachlan, A. (2008). Sandy beach ecosystems: key features, sampling issues, management challenges and climate change impacts. *Marine Ecology* 29, 70–90.

Service, R. F. (2006). Desalination freshens up. *Science* 3113, 1088–1090.

Smith, M. D., Slott, J. M., McNamara, D. & Murray, A. B. (2009). Beach nourishment as a dynamic capital accumulation problem. *Journal of Environmental Economics and Management* 58, 58–71. DOI: 10.1016/j.jeem.2008.07.011.

Smyder, E. A. & Martin, K. L. M. (2002). Temperature effects on egg survival and hatching during the extended incubation period of the California grunion, *Leuresthes tenuis*. *Copeia* 2002, 313–320.

Spratt, J. D. (1986). The amazing grunion. *Marine Resource Leaflet* 3. Sacramento: California Department of Fish and Game.

Takasuka, A., Oozeki, Y. & Aoki, I. (2007). Optimal growth temperature hypothesis: why do anchovy flourish and sardine collapse or vice versa under the same ocean regime? *Canadian Journal of Fisheries and Aquatic Sciences* 64, 5, 768–776. DOI: 10.1139/f07-052.

Tang, S. & Chen, L. (2004). The effect of seasonal harvesting on stage-structured population models. *J. Math. Biol.* 48, 357–374.

Valentine, D. W. & Soulé, M. (1973). Effect of *p'*-DDT on developmental stability of pectoral fin rays in the grunion, *Leuresthes tenuis*. *Fishery Bulletin* 71, 920–921.

Winkler, D. L., Duncan, K. L., Hose, J. E. & Puffer, H. W. (1983). Effects of benzo(a) pyrene on the early development of the California grunion, *Leuresthes tenuis* (Pisces, Atherinidae). *U.S. National Marine Fisheries Services, Fishery Bulletin* 41, 3, 473–481.

Yates, M. L., Guza, R. T., O'Reilly, W. C. & Seymour, R. J. (2009). Overview of seasonal sand level changes on Southern California beaches. *Shore & Beach* 77, 39–46.

Zhang, K., Douglas, B. C. & Leatherman, S. P. (2004). Global warming and coastal erosion. *Climate Change* 64, 41–58.

10 Waves of Passion: Conservation Efforts for Beach-Spawning Fishes

It is nearly impossible today for any type of wildlife to find a pristine location absent from human influence. The same can be said for many beach-spawning fishes that come to the water's edge along heavily populated coastlines, where their vulnerable embryos develop in the footprints of cities, ports, marinas, and recreation areas. How can these small, innocent fishes manage to survive in the face of all these threats? Whether by direct harm or simply by neglect, the anthropogenic impact is large, and the result for the fishes is often negative.

Safe nursery areas on the coast are critical for the reproduction and early life stages of beach-spawning fishes, and vital to the survival of the species. In Chapter 9 some of the current threats to beach habitats are detailed. These dangers are global and will continue to increase as more beach habitat is lost to construction or shoreline armoring along the coastlines of the world.

Species of beach-spawning fishes occur worldwide, on all continents except Antarctica, and there are undoubtedly many additional species that spawn on beaches that have not yet been described. Protecting them from the many threats to their critical reproductive habitat requires favorable policies and positive public opinion. Protecting wildlife requires collaboration between the public, resource management agencies, and scientists. These complex threats are addressed by protections and enforcement of regulations involving multiple stakeholders, sometimes with competing or conflicting purposes.

Grassroots conservation programs to protect beach-spawning fishes are highlighted in this chapter. Fortunately, several of these beach-spawning fishes have attracted the admiration of human supporters. These people volunteer their time to actively monitor the runs and communicate with scientists and resource managers about their concerns. These "a-fish-ionados" may be loosely organized or independent, local or spread over a large area, trained citizen scientists or simply intrigued observers. Each remarkable effort succeeds only with determination, persistence, and initiative.

10.1 CALIFORNIA GREETS THE GRUNION

One charismatic beach-spawning fish is the California grunion. This species has such unusual spawning behavior that people do not believe it until they see it. See Chapter 4 for details. The first published scientific description of these fish out of water was a fisherman's tale, reported with a bit of skepticism and the admonishment, "A detailed study of these interesting habits, or a confirmation of them, is highly desirable" (Hubbs, 1916, p. 161).

Public programs to observe grunion runs started in 1949 at Cabrillo Beach in San Pedro, led by the always affable John Olguin, a career ocean lifeguard, waterman, and observant student of marine biology. The popularity of his program was evident from the start, when cars backed up for miles to get onto the beach on program nights (Harzen et al., 2011). This public outreach continues today, and hundreds of thousands of visitors have shared the thrill of the grunion run at Cabrillo Marine Aquarium over the years. The aquarium incorporates the grunion into their logo (Figure 10.1) and sets aside one night a year for their "Fish-tival" to celebrate all things grunion. To top it off, their annual formal fundraiser is called the "Grand Grunion Gala."

Because California grunion runs are synchronized with the tides, they are fairly predictable (Walker, 1952). This makes it possible for many different organizations along the coast of California to offer public programs on summer nights, including the Birch Aquarium at Scripps in La Jolla, the Ocean Institute in Dana Point, the Crystal Cove Alliance, Bolsa Chica State Park, the Doheny State Beach Interpretive Association, and others.

California grunion are endemic to the Pacific coast of California and Baja California; that is, they are found nowhere else. The spawning sites for this fish are the beautiful sandy beaches along the coast of Southern California. These coastal strands are famous recreational areas for sun, sand, and surf and attract millions of people every year and contribute billions of dollars to the state's economy. Many public beaches are cleaned by raking and shaping the sand, using heavy equipment such as loaders, tractors, and graders (Martin et al., 2006). On some popular destinations for recreation, beaches are raked daily. This disturbs the sand surface and any living thing that may be present in the upper layers of sand (Dugan et al., 2003). See Chapter 9 for more on these effects.

One individual in San Diego, California, developed a concern over the potential harm that beach grooming could be doing to grunion eggs buried in the sand. On a walk one morning, she saw birds swooping down to feed in the tracks formed behind one of the loaders that was raking the shore. To her horror, she saw the birds consuming bright orange grunion eggs turned up by the tines. A year later, unable to spot any grunion running on her local beach. Pat Gallagher took her concerns to the city council. She got their rapt attention by stating, "There are no grunion left in San Diego! Beach grooming has killed them all."

FIGURE 10.1   The logo for Cabrillo Marine Aquarium in California features a California grunion surfing toward shore on a moonlit wave. (Reprinted with permission.)

The city decided to look into this matter by requesting the formation of a Blue Ribbon Panel to examine the situation. A local nonprofit environmental group, Project Pacific, took on the challenge. Their president, Melissa Studer, recruited experts from the National Marine Fisheries Service, the California Department of Fish and Wildlife (CDFW, then called California Department of Fish and Game), the U.S. Fish and Wildlife Service, the city's Shoreline Parks department, the local Audubon Society, two distinguished scientists from nearby Scripps Institution of Oceanography, and others. It was decided that a scientific study was necessary to answer two questions: Were any grunion still spawning in San Diego, and if so, was beach raking harmful to their nests?

Both ichthyologists from Scripps, Jeffrey Graham and Richard Rosenblatt, are brilliant, knowledgeable scientists, but neither had ever done research on the California grunion. When this author offered to provide advice because of previous research with the eggs and embryos of California grunion, they extended an invitation to design and carry out the study.

The first quest was to discover whether grunion still ran on local beaches. A brief preliminary study in 2001 was not very encouraging. Several observers saw no fish running at all during several nights of July and August. The timing for grunion runs is fairly predictable, following the extensive dissertation research by Boyd Walker (1949), and the dates of predicted runs are published every year by the CDFW. However, in the more than 50 years since Walker's initial study, an error had crept into the calculations for the timing of the nights when runs are most likely to occur, shifting it by two days. People trying to observe the runs were frustrated, in part because they were going on some of the wrong nights. In addition, July and August are the least likely time to see large runs. The outpouring of public concern for this species encouraged development of a method for volunteers to make observations on several city beaches.

In previous discussion with Mike Schaadt and Susanne Lawrenz-Miller, the three of us produced a rubric for scoring the strength and duration of grunion runs (Martin et al., 2007) so as to make long-term comparisons between Malibu and San Pedro beaches (see Table 10.1). A similar system was previously developed for scoring the runs of the Atlantic silverside, *Menidia menidia* (Middaugh, 1981). The Walker Scale scoring system was augmented for data acquisition to include additional information about the weather conditions, predator presence, and human activities during the time of the predicted run, for use by volunteers from the public.

In 2002 the first training workshops were held for volunteer citizen scientists, whom we dubbed Grunion Greeters, at Birch Aquarium at Scripps. More than 150 people turned out that first year to learn this method for making observations. They were provided with precise schedules and specific locations for observations on the nights and at the times that runs were most likely to occur. At the appointed dates, these intrepid volunteers made their way late at night through the darkness in hopes of spotting some small silver fish that might or might not appear out of the waves.

Happily for everyone, California grunion showed up at the appointed times and places on city beaches in San Diego (Figure 10.2). Subsequent experimental

TABLE 10.1

Assessment of the Strength of Spawning Runs of California Grunion,
Leuresthes tenuis* (Atherinopsidae) by Use of the Walker Scale

Scale	Numbers of Grunion	Descriptor
W-0	No fish on beach or only a few individuals, no spawning, no "sticking"	not a run
W-1	10 to 100 scattered on beach at a time, some spawning, some fish "stick" on beach between waves	light run
W-2	100 to 500 spawning on several areas of beach at different times	moderate run
W-3	Hundreds of fish spawning in several areas of beach at peak of run, or repeatedly over time in one area	good run
W-4	Hundreds of fish on beach close together, little sand visible between fish at peak of run; peak lasts less than an hour	strong run
W-5	Fish covering the beach several individuals deep along the entire length of the beach at peak of run; a silver lining along the surf; peak lasts more than an hour	incredible run (less than 2% of all reports)

* Assessed by numbers of fishes on shore, duration of run, and extent of shore involved in the run.

FIGURE 10.2 (See color insert.) Grunion Greeters find their target species on California beaches. (Photo by Jennifer Flannery Harr. With permission.)

manipulation showed that raking over the areas where runs had been observed was indeed harmful to the eggs (see Chapter 9 for more on this). With the support of the beach manager, Dennis Simmons, and the San Diego City Council, beach mainte-nance protocols were changed to protect these nursery areas during the spawning season (Martin et al., 2006).

Over the next decade public enthusiasm spread, and the Grunion Greeter program was expanded across the state of California and into Mexico, throughout the habitat range of this endemic species (Martin et al., 2007). Hundreds of volunteers were trained each year at multiple locations along the coast. Volunteers discovered a northward range extension (Roberts et al., 2007) and followed a habitat expansion into San Francisco Bay (Johnson et al., 2009; Martin et al., 2013), and other coastal communities were persuaded to alter their beach maintenance procedures to protect the California grunion nests (Martin et al., 2011). New policies for protection of grunion spawning areas are now included in coastal construction permits.

From the beginning, grunion spawning-run data were collected online via a web-based questionnaire and instantly available to scientists. Participants can also call a telephone hotline to report large runs or any concerns about the beach-spawning sites. Frequent communication was maintained between volunteers and the scien-tific team via e-mail and web postings. The Grunion Greeters trained more than 4000 volunteers in a decade, involved more than 140 environmental organizations, and received hundreds of data reports every year. See Table 10.2 for a list of the com-munity partners and www.grunion.org for more on this group.

Maps of beaches where *Leuresthes tenuis* is known to spawn were created from reports from the Grunion Greeters and historical data. They are available as a layer in MarineMap, used in the statewide planning of the system of Marine Protected Areas for California (California Marine Life Protection Act Initiative, 2009).

California grunion are protected from fishing during a closed season in April and May, but spawning fishes may be harvested with no bag limit at other times. Thousands of people waiting onshore may far outnumber the grunion in the waves, and every fish that approaches may be taken. See Chapters 6 and 9 for more on this.

A new nonprofit organization, the Beach Ecology Coalition, grew out of the Grunion Greeters because of the frequent interactions between beach managers, ecologists, environmental activists, lifeguards, and surfers from across Southern California. These ocean stakeholders had not met together or interacted on a regu-lar basis in the past. This group strives to incorporate ecologically sensitive prac-tices into beach and coastal management for all the animals and plants that depend on beaches and to educate the public about the ecological functions of beaches (http://www.beachecologycoalition.org).

Increased public awareness has extended to government agencies that manage natural resources along the coast, including the California Coastal Commission, California State Parks, and the U.S. Army Corps of Engineers. Recognition and appreciation for this charismatic natural phenomenon has dramatically improved the protection of grunion spawning areas and has increased awareness of and protection for other beach denizens along the way. Projects for coastal construc-tion, beach sand replenishment, shoreline armoring, and beach grading are all subject to ecological evaluation for impacts on grunion runs. Soon a new project

TABLE 10.2
List of More than 150 Cooperating Organizations That Assisted with Local Community Involvement of Volunteers and Implementation of Findings with the Grunion Greeter Project in the Past Decade

Region	Cooperating Organization
San Diego County	Scripps Institution of Oceanography; Birch Aquarium at Scripps; City of San Diego; Buena Vista Audubon Society; City of Oceanside; Camp Pendleton Marine Corps Station; Palomar College; MiraCosta College; City of Imperial Beach; Tijuana River Estuarine Research Reserve; Southwest Wetlands Interpretive Association; Surfrider Foundation, San Diego Chapter; City of Imperial Beach; City of Solana Beach; University of California, San Diego; San Diego State University; Coronado Naval Air Base; SANDDAG; San Elijo State Beach Interpretive Association; Project Pacific; San Diego Science Festival; Volunteer San Diego; Grunion Festival Kumeyaay Tribe
Orange County	Surfrider Foundation National Headquarters and Orange County Chapter; Muth Interpretive Center, Newport Bay; City of Newport Beach; Ocean Institute at Dana Point; City of San Clemente; Crystal Cove Alliance; California State University, Fullerton; Southern California Ecosystems Research Program; Doheny State Beach Interpretive Association; Bolsa Chica State Beach Interpretive Association; City of Laguna Beach; Orange County Coastal Coalition; County of Orange; City of Huntington Beach; City of Seal Beach; Whittier College
Los Angeles County	Aquarium of the Pacific in Long Beach; Cabrillo Marine Aquarium; Heal the Bay; Santa Monica Pier Aquarium; Roundhouse Aquarium at Manhattan Beach; City of Long Beach; Los Angeles County Beaches and Harbors; Los Angeles County Fire Department, Lifeguard Division; Pepperdine University; University of California, Los Angeles; Malibu Watershed Council; GLOW art fair; Santa Monica; Santa Monica Bay Restoration Commission; Santa Monica Baykeepers; California State University, Northridge; City of Malibu; California State Parks Whale Fiesta; Marina Del Rey Marine Science Academy; Surfrider Foundation, West LA–Malibu Chapter; "Chance for Children" program for at-risk children; City of Santa Monica; LA Audubon Society; Los Angeles County Natural History Museum; Occidental College; Santa Monica Bay Audubon Society, Pepperdine University; University of Southern California; Pomona College; Los Angeles County Beach Commission; Las Virgenes Resource Conservation District; Santa Monica Mountains Preservation Trust; LA Works; Grunion.org
Ventura County	Oxnard City College; Ormond Beach Task Force; City of Oxnard; California State University, Channel Islands; Ventura County Coastal Coalition
Santa Barbara County	Santa Barbara Channel Keepers; Santa Barbara Natural History Museum; Ty Warner Sea Center; University of California, Santa Barbara; City of Santa Barbara; County of Santa Barbara; Goleta Beach Task Force
San Luis Obispo County	Oceano Dunes Natural Recreation Area

TABLE 10.2 (*Continued*)
List of More than 150 Cooperating Organizations That Assisted with Local Community Involvement of Volunteers and Implementation of Findings with the Grunion Greeter Project in the Past Decade

Region	Cooperating Organization
Monterey County	Pacific Grove Museum; Monterey Bay Aquarium Research Institute; Moss Landing Marine Laboratory; University of California, Santa Cruz
San Francisco Bay Area	Port of Oakland; East Bay Regional Parks District; Golden Gate National Recreation Area; Crissy Field Interpretive Association; Cordell Banks National Marine Sanctuary; Pt. Reyes National Seashore; Lawson's Landing Campground; City of Marshall; City of Oakland; City of Alameda; City of San Leandro; University of California, Berkeley; Stanford University
Statewide, California	California State Parks; California Coastal Commission; California Department of Fish and Wildlife; California Marine Life Protection Act; California Sea Grant College; California Sediment Management Work Group; Beach Ecology Coalition; California Coastal Coalition; Clean Beaches Task Force; California Waterboards; Southern California Academy of Sciences
National/International	National Marine Fisheries Service; National Oceanic and Atmospheric Administration; National Science Foundation; National Fish and Wildlife Foundation; Environmental Protection Agency; Oil Spill Preparedness and Response Team; National Geographic Society; National Wildlife Federation; Wildlife Watch Week; Young Round Square of the Americas Conference; Calidad De Vida; Tijuana, Mexico; No Bad Days Tourism; San Felipe, Mexico; US National Park Service; US Army Corps of Engineers; Shoreline Preservation Association; Citizen Scientist Toolkit Conference; NatureServe National Phenology Network; American Institute of Fishery Research Biologists; American Fisheries Society; American Shore and Beach Preservation Association

is will start to monitor the ecological components of beaches, with the assistance of local community volunteers. This program, All Ashore, continues the legacy of the Grunion Greeters (http://www.allashore.org).

10.2   BEACH-SPAWNING FISHES HAVE FRIENDS ON THE SHORELINE IN WASHINGTON STATE

In the state of Washington, no splashy midnight spawning runs entertain crowds, but several species of fishes reproduce on gravel beaches within Puget Sound. These include surf smelt (*Hypomesus pretiosus*) and Pacific sand lance (*Ammodytes hexapterus*) (Penttila 1995a, 2007). These small forage fishes are more important for their ecological role than their economic value in a direct fishery. Surf smelt spawn year-round in some areas of Puget Sound. Fishing on the spawning fish is allowed along two miles of shoreline, and several tons of them are taken each year. But over a huge area of the coast, surf smelt are not fished, in part because of beach privatization. There are no legally designated refuges on the Puget Sound shoreline for beach-spawning fish. Pacific sand lance are not fished at all (Penttila, 2007). Like many of the beach-spawning fishes, surf smelt and Pacific sand lance are important in the marine food web, connecting their prey, the primary planktonic production to their predators, the higher trophic levels of larger fish, particularly salmon, seabirds, and marine mammals (Pikitch et al., 2014).

In some places within Puget Sound, both surf smelt and sand lance spawn on the same beaches but at different tidal heights (see Chapter 7). In winter, occasionally the eggs of a third species, rock sole (*Lepidopsetta bilineata*), may also co-occur on some beaches (Penttila, 1995b). The spawning runs of these species are rarely observed; Pacific sand lance runs were first observed only within the past decade by amateur naturalists, and the flatfish runs have not yet been observed. Their small, adherent eggs attach to gravel and are tidally exposed to air.

Additional forage fishes spawn near shore in Washington State. Pacific sandfish *Trichodon trichodon* are important as forage fishes for some predators such as the Steller sea lion *Eumetopias jubatus*. Pacific sandfish depend on nearshore areas for spawning, egg incubation, and larval stages (Thedinga et al., 2006). Pacific herring *Clupea harengus pallasi* spawn on surfgrass and other complex biogenic habitats near shore (see Chapter 4 for details).

Forty years ago, there were no maps of spawning grounds for forage fishes and no regulatory habitat protection. Since 1972, a small group of dedicated volunteers and Washington Department of Fish and Wildlife (WDFW) staff biologists have put boots on the ground and walked the beaches to gather gravel by the bucketful to locate the spawning grounds of these forage fishes.

According to Dan Penttila (personal communication), "Public education and involvement are key" to the protection of beach-spawning fishes. In 2001, the Northwest Straits Commission provided funding to promote citizen science for seven counties in Puget Sound (Moulton & Penttila, 2001). This provided the impetus for a group called the Beach Watchers in an effort to map the spawning habitats of surf smelt and Pacific sand lance throughout Puget Sound (Quinn et al., 2012). Volunteers received extensive training and instructions at the "Shoreline Institute." They learned how to sample 100 feet of shoreline with perpendicular transects between the low and high tide lines (Moulton & Penttila, 2001). Samples of gravel from beaches with likely habitat were collected and washed into buckets to remove the tiny, nearly invisible eggs. The samples were preserved and sent to biologists for identification and analysis at the WDFW (Penttila, personal communication). Similar efforts were undertaken on Vancouver Island and sent for processing to Bamfield Marine Station. Table 10.3 lists many of the organizations involved in this effort around the Salish Sea.

Public involvement of the Beach Watchers went beyond searching for essential fish habitat with shoreline transects. Some groups organized nature walks for the public or monitored for seabird carcasses on beaches; others worked to promote dam removal or other types of habitat restoration.

The spawning habitats of surf smelt, Pacific sand lance, sandfish, herring, and rock sole are considered marine habitats of special concern by the Washington Administrative Hydraulic Code. Over the course of a decade, policies developed by WDFW became regulatory language for the Washington State codes. Initial applications of these policies and new rules for coastal permits were met with consternation but led to public hearings and trials that built up a body of evidence over the years that became codified

TABLE 10.3

List of Organizations in Salish Sea Working with Volunteers for Forage Fishes in Washington State and in British Columbia, Canada

Northwest Straits Commission, based at Padilla Bay NERR, Bayview, WA, www.nwstraits.org
  Contact: Caroline Gibson, gibson@nwstraits.org

Northwest Straits Foundation, based in Bellingham, WA, www.nwstraits.org
  Contact: Joan Drinkwin, drinkwin@nwstraits.org

Friends of the San Juans, based in Friday Harbor, WA, www.sanjuans.org
  Contact: Tina Whitman, tina@sanjuans.org

Island Co. Marine Resources Committee/Island Co. Beachwatchers, based in Coupeville, WA
  Contact: Sarah Schmidt, 4bats@ixoreus.com

Coastal Watershed Institute, based in Port Angeles, WA, www.coastalwatershedinstitute.org
  Contact: Anne Shaffer, anne.shaffer@coastalwatershedinstitute.org

Snohomish Co. Marine Resources Committee, based in Everett, WA, www.snocomrc.org
  Contact: Kathleen Herrmann, kathleen.herrmann@co.snohomish.wa.us

Snohomish Co. Beachwatchers, based in Everett, WA, www.beachwatchers.wsu.edu/snohomish
  Contact: Chrys Bertolotto, chrys@wsu.org

Friends of Skagit Beaches, based in Anacortes, WA, www.skagitbeaches.org
  Contact: Nancy Olsen, nancy.skagitbeaches@gmail.org

British Columbia Shore Spawners Alliance, based in Mill Bay, BC
  Contact: Ramona de Graaf, foragefish.bc@gmail.com

legally in the State Shoreline Management Act. This act contains specific language for "no net loss" of spawning habitat for forage fishes, and it states that all documented spawning sites are considered essential fish habitat. See Figure 10.3 for a list of the saltwater habitats of special concern that are protected by the Hydraulic Code. However, in spite of these protections, single families may still apply for permits for shoreline bulkheads, and the WDFW is not allowed to refuse these permits.

Beaches are affected by many types of coastal construction. In Puget Sound, boat ramps, sea walls, culverts, dikes, riprap, and bulkheads alter long-term transport of sediment or bury spawning habitat. An organization on the San Juan Islands in northern Puget Sound, the Friends of the San Juans, has suggested the creation of "beach protection/no build zones" for the 9 percent of shoreline that is documented to have forage fish spawning habitats and feeder bluffs in the islands (D. Penttila, personal communication).

South of Everett, Washington, an area of Puget Sound shoreline is bordered by the Burlington–Northern/Santa Fe railroad causeway, originally positioned directly upon the upper intertidal beach in the late 1880s. The Snohomish County Marine Resources Committee has proposed to restore several thousand feet of upper intertidal beach, including spawning habitat for surf smelt and Pacific sandlance, to a more productive condition through application of beach sediment. A previous attempt to restore a beach for spawning of surf smelt was not particularly effective because that area was prone to lateral transport and the new substrate material drifted away (D. Penttila, personal communication). However, restoration of habitat by removal of construction may facilitate natural replenishment of sediments over time.

"In the following saltwater habitats of special concern, or areas in close proximity with similar bed materials, specific restrictions regarding project type, design, location, and timing may apply as referenced in WAC 220-110-270 through 220-110-330. The location of such habitats may be determined by a site visit. In addition, the department may consider all available information regarding the location of the following habitats of special concern.

(1) Information concerning the location of the following saltwater habitats of special concern is available on request to the habitat management division of the department of fish and wildlife. These habitats of special concern may occur in the following types of areas:

(a) Surf smelt (*Hypomesus pretiosus*) spawning beds are located in the upper beach area in saltwater areas containing sand and/or gravel bed materials

(b) Pacific sand lance (*Ammodytes hexapterus*) spawning beds are located in the upper beach area in saltwater areas containing sand and/or gravel bed materials.

(c) Rock sole (*Lepidopsetta bilineata*) spawning beds are located in the upper and middle beach area in saltwater areas containing sand and/or gravel bed materials.

(d) Pacific herring (*Clupea harengus pallasi*) spawning beds occur in lower beach areas and shallow subtidal areas in saltwater areas. These beds include eelgrass (*Zostera* spp) and other saltwater vegetation and/or other bed materials such as subtidal worm tubes.

(e) Rockfish (*Sebastes* spp) settlement and nursery areas are located in kelp beds, eelgrass (*Zostera* spp) beds, other saltwater vegetation, and other bed materials.

(f) Lingcod (*Ophiodon elongatus*) settlement and nursery areas are located in beach and subtidal areas with sand, eelgrass (*Zostera* spp), subtidal worm tubes, and other bed materials.

(2) Juvenile salmonid (Family Salmonidae) migration corridors, and rearing and feeding areas are ubiquitous throughout shallow nearshore saltwater areas of the state.

(3) The following vegetation is found in many saltwater areas and serves essential functions in the developmental life history of fish or shellfish:

(a) Eelgrass (*Zostera* spp);

(b) Kelp (Order Laminariales);

(c) Intertidal wetland vascular plants (except noxious weeds)."

[Statutory Authority: RCW 75.08.080. 94-23-058 (Order 94-160), § 220-110-250, filed 11/14/94, effective 12/15/94. Statutory Authority: RCW 75.08.012, 75.08.080 and 75.20.100. 84-04-047 (Order 84-04), § 220-110-250, filed 1/30/84. Statutory Authority: RCW 75.20.100 and 75.08.080. 83-09-019 (Order 83-25), § 220-110-250, filed 4/13/83.

FIGURE 10.3 Washington State Code 220-110-250, saltwater habitats of special concern.

10.3 CANADIANS CARE ABOUT CAPELIN

The capelin roll is cause for celebration in Canada. In the past, feasts were held in the St. Lawrence coast and cities organized festivals celebrating the arrival of these spawning fishes (Boulay, 2011). Today, instant communication allows lucky observers to alert their neighbors of the arrival of capelin *Mallotus villosus* on a local beach.

Along the Atlantic shores of Canada, a Capelin Observers Network (CON) has operated since 2003 (Boulay, 2011). This was established for the public by the Oceans Management Division of Fisheries and Oceans Canada to track locations of spawning sites and, to some degree, the strength or extent of the spawning runs of the capelin. Originally developed for the St. Lawrence area, it has since expanded to include capelin spawning sites all along the Atlantic coast of Canada. Historical data since 1945 is also incorporated into their maps.

As capelin runs are not strictly predictable from tides, these observations are opportunistic and depend on people being at the right place at the right time (Templeman, 1948). Then they must provide that information to the CON. One signal that the capelin run is beginning is the strong smell of cucumber that emanates from these smelt, due to the chemicals in their skin (Love, 2011).

Information can be reported online, by telephone, or via e-mail. While offshore, capelin are fished commercially, but onshore during spawning runs, capelin are a recreational fishery (see Chapter 6). CON suggests assessment of the strength of the runs by recording the number of minutes it takes to catch a 20-liter bucket of fish, an example of catch per unit effort. Runs can also be classified as high, average, or low; but not all runs are reported this way. Observation locations are mapped, and these locations are available online to anyone. Annual reports indicate the substrates used by spawning capelin, the relative strength of runs reported, and the timing of the runs.

These observations reveal a surprising variability in the timing of peak runs in different months across a rather limited geographic area. Reports also indicate a greater dependence on sandy beaches in the St. Lawrence area, more so than gravel beaches that are commonly seen as spawning beaches in Newfoundland and Labrador. Subtidal spawning is not monitored by this effort (Nakashima & Wheeler, 2002). However, one observer wrote: "It was the festival of capelin in Marsoui between 21 and 23 of June 2001. It spawned on the beach but also a lot on the bottom in 5–6 meters of water" (Frédéric Hartog, July 2011, quoted in Boulay, 2011).

Unfortunately there is not a cohesive support group for these observers, and the number of observers varies greatly from year to year, from as low as 19 and up to 150. Observers in Newfoundland and Labrador report little data to the CON. The variability in effort and observers makes it difficult to assess the capelin population status in general or even on the local beaches from one season to the next (Smith & Michels, 2006).

Additional information is gathered by telephone surveys of anglers (Brian Nakashima, personal communication). Management by Fisheries and Oceans Canada takes the position that the recreational and commercial fisheries on capelin take only a small proportion of the total population and do not impact the species negatively (Boulay, 2011). The agency does however state its intent to protect the beach-spawning sites of this ecologically and commercially important species from development or other forms of human disturbances. See Chapter 9 for more on the commercial fishery for capelin.

10.4   KUSAFUGU ARE CULTURAL TREASURES IN JAPAN

Of the 25 species within the puffer genus *Takifugu*, only one spawns on beaches, *Takifugu niphobles* (Uno, 1955). In Yamaguchi Prefecture, Japan, the beach where these puffers come to spawn was designated as a natural monument in 1969. *Tsutsumigaura Beach in Hikari City* is the protected location for kusafugu puffer observation parties during the spawning season (*Hikari City Lifelong Learning website*). Many locals attend to watch and collect data on the time of spawning, the number of fish, and other conditions (Figure 10.4). No fishing or take is allowed during the runs. Scientific collectors must obtain permission from the city authorities before working with this species (Kazunori Yamahira, personal communication),

FIGURE 10.4 Visiting the kusafugu puffer run in Hikari City, Japan. (Photo used by permission of Hikari City Board of Education, Culture, and Lifelong Learning Division, http://www.city.hikari.lg.jp/bunka/kusahugu25.html.)

FIGURE 10.5 (See color insert.)   The spawning run of the kusafugu puffer. (Photo used by permission of Hikari City Board of Education, Culture, and Lifelong Learning Division, http://www.city.hikari.lg.jp/bunka/kusahugu25.html.)

and the police provide the fish with protection from unauthorized take of this natural treasure. The presence of a spawning area on the coast provides added incentive for the citizens to prevent coastal pollution.

The city Board of Education, Culture, and Lifelong Learning Division, Cultural Promotion Engagement invites the public and the press to join teams of observers. People attending the runs are advised about the kinds of behavior to expect. They are asked to wait quietly, avoid disturbing the spawning fishes, and turn off the flash on their cameras. Because the runs occur close to shore during the late afternoon, wonderful photos can result (Figure 10.5).

10.5 CELEBRATIONS OF BEACH-SPAWNING FISHES IN SAUSALITO, CALIFORNIA

Richardson Bay, near Sausalito, California, is one spawning site for a large contingent of plainfin midshipmen. Male plainfin midshipman, *Porichthys notatus,* set up nesting sites under boulders in the intertidal zone and then create humming vibrations of the swim bladder to attract females to breed (see Chapter 4 for more on this). This odd noise may generate only mild curiosity from humans in remote locations (Love, 2011), but in a bay filled with houseboats, the noise generated by the singing fish is so impressive that some who have heard it suggested it was caused by a spying submarine or some other underwater machine (Sutter, 2012).

John McCosker of San Francisco's Steinhart Aquarium identified the chorus as the amorous sounds of many piscine suitors in their bachelor pads. Now, every July, the return of the rumbling sound is greeted with some fondness and romanticism. Some folks make comparisons to the swallows' annual return to San Juan Capistrano.

Residents vary in their appreciation for the fishes, and in fact preposterous explanations for the summertime sound still abound. "'Perhaps it is an orchestra of toadfish,' said doubter Hugh Lawrence. 'When you show me the leader in his little tuxedo and baton lining up the chorus and starting them all at the same time, I'll believe'" (Sutter, 2012). Recently, during an Independence Day parade, "Atop a float … sat Michele Affronte, dressed as a mermaid with a sparkly blue tail and a crown of wooden sea stars. She said the float was made in honor of the humming toadfish, which often gets killed while feeding off the bottom of boats. 'I am the humming toadfish mermaid queen'" (Hansen, 2013). (Reports of toadfish getting killed while feeding off boats are nonexistent. However, some of the males die after their childcare duties are completed. See Feder et al., 1974.) One anonymous local provided this mangled information: "It's the plainfinned midshipman, as I understand, with the phosphorescent genitals that vibrate, okay" (interview by Bryant Gumbel on *NBC Today Show,* 1986).

Charming (or alarming) as these anecdotes are, there does not seem to be any additional effort by the local citizens to protect spawning areas or to celebrate this unique species. On the plus side, because of the unappealing appearance of *P. notatus* there is also no interest in a recreational fishery on this species. A conservation effort on behalf of this particularly quirky creature might be an opportunity to protect spawning locations, as a way to show civic pride and local color in an area that is known for its ecological awareness.

Recently, people at the Cass Gidley Marina in Richardson Bay have begun promoting a Sausalito Herring Festival. The one-day event includes gourmet food, entertainment, and herring-themed arts and crafts. The program includes information about the fishery in San Francisco Bay for herring, as well as education about the ecological role of the herring run for predators such as seabirds, marine mammals, and other fishes. As the last remaining commercial fishery in San Francisco Bay, the herring fishery is carried out by visitors but the locals are concerned with environmental sustainability and with maintaining its economic viability. The combination of economic and environmental goals makes this a noteworthy effort.

10.6   WHITEBAITERS IN NEW ZEALAND PROTECT THEIR RECREATION

New Zealand is a relatively new country with a small population. Although its Department of Conservation has broad powers to protect and conserve the endemic native flora and fauna, the whitebait fishery is unique in having few regulations and little scientific basis for this consumptive use (McDowall, 1996). This fishery targets the small juvenile fishes rather than the adults, so they are taken before they have matured to reproduce. No license is required to catch the incoming juvenile fishes as they leave the ocean to swim upstream in fall; no limits are placed on the amount anyone can catch, and there is no license required to sell the catch, which can reach a price as high as $150NZD per kilogram (http://www.teara.govt.nz/en). Regulations limit the open season and the types of gear that can be used, and limit each angler to only one net at a time that must be personally attended. In 1992 news that stricter regulations were being considered inspired the formation of a new organization, the Southland Recreational Whitebaiters Association (SRWA). This group united anglers hoping to protect their fishing interests, keep numerous privately owned fixed fishing stands, and continue to access fishing sites along rivers (Haggerty, 2007). The SRWA succeeded in defending their fishing activities but later, because of their keen interest, became involved in addressing some biological questions and developing policies for conservation. These included identifying previously unknown spawning sites and increasing awareness of river pollution and its effects on the anadromous Galaxiid fishes that make up the whitebait fishery (McDowall, 1996). With increased involvement in policy development, members of SRWA also took on some of the enforcement of regulations when they observed poachers or anglers out of season (Haggerty, 2007). Restrictions on net placement and size provide some protection, along with closed seasons, but no bag limits have been implemented yet.

The involvement of recreational anglers or hunters in conservation activities is not unique, but for the members of SRWA to move from an almost unregulated system of harvest to an active involvement in habitat conservation and resource management is impressive. However, four of the five Galaxiid fishes caught as whitebait are considered "declining," including the inanga, *Galaxias maculatus*, the most commonly caught species that comprises approximately 95% of the take (Whitebait Brochure, 2011). Population numbers are unknown and catch reports are unreliable (Haggerty, 2007). Still, current conservation efforts involve stabilizing stream edges, a closed season, and gear restrictions, rather than placing limits on the harvest. The introduced brown trout has been a strong predator on these small forage fishes in New Zealand, but McDowall (2006, p. 269) suggests that the "huge numbers caught in the whitebait fishery" are more influential than predation. Ironically, the introduced trout may enjoy greater conservation status than these small native fishes.

Europeans have a long tradition of fishing for whitebait, which in that part of the world includes juvenile herring, sprat, goby, shad, and sand eel, commonly served fried with lemon. The Marine Conservation Society UK considers whitebait "the least sustainable type of fish to eat" (http://www.fishonline.org/fish/whitebait-285), and has recommended avoidance of consumption of whitebait because of the ecological consequences of removing juvenile forage fishes from the oceanic food web before they breed (Smith, 2012).

10.7 FORAGE FISH INITIATIVES ADDRESS ECOLOGICAL CONCERNS

Ecological concerns about the oceanic food web drive the Pacific Fish Conservation Campaign of the Pew Charitable Trusts (http://www.pewenvironment.org). This program grew out of concern over the potential for new fisheries targeting previously unfished species, a consequence of depletion of stocks of larger, more traditionally fished species (Myers & Worm, 2003; Pauly et al., 1998). These smaller fishes link the primary productivity of marine microalgae to the larger fishes, seabirds, and marine mammals. Removing them can alter the ability of species in the higher trophic levels to recover from overexploitation (Frank et al., 2011, 2013; Vandepeer & Methven, 2007), but different life cycle stages of each species may alternate between predator and prey, creating a complex management dynamic (Hjerrmann et al., 2004). The Pacific Fishery Management Council (PFMC) is a government organization that manages ocean fisheries of California, Oregon, and Washington. As part of its efforts at ecosystem-based management, PFMC is considering an unfished and unmanaged forage fish initiative that would require substantial biological data before a new fishery could be started on a previously unfished species, particularly one that has not previously been managed (http://www.pcouncil.org). The Lenfest Ocean Program also intends to increase protection of some forage fishes. This foundation has named a Fishery Ecosystem Task Force to develop recommendations for ecosystem-based management of large marine ecosystems (http://www.lenfestocean.org).

Some of the beach-spawning fishes have long been commercially harvested, including capelin and Pacific herring. Others are the focus of recreational fisheries that can be very intense, including whitebait, California grunion, and surf smelt. Besides their human consumers, all of these species have natural marine predators. Changes from either the top or the bottom of the marine food web will have consequences that are difficult to predict (Pikitch, 2010; Pikitch et al., 2014). Choosing a new species to target for fishing should be approached with caution, particularly as these populations fluctuate with oceanographic conditions and climate change (Martin et al., 2013; Robards et al., 2002; Rose, 2005). At the same time, as all anglers know, it is also vital to protect spawning habitat, especially for those fishes that spawn along the shores of human habitations.

10.8 SUMMARY: LOCAL ECOLOGICAL KNOWLEDGE IS NEEDED FOR CONSERVATION

Many conservation organizations rely on charismatic wildlife to attract participants and appeal to the public. The World Wildlife Fund has their panda, and the Galapagos Conservancy has the giant tortoise Lonesome George. Salmon and trout inspire conservation for their recreational value, but most fish species are unlikely to inspire organized grassroots efforts to cherish and protect them. It is heartwarming to see some beach-spawning fishes draw the attention and concern of individuals who then develop groups that endeavor to protect these amazing creatures. This is all the more remarkable when these species are not exploited for commercial gain or are endemic to a local area.

The conditions that threaten the survival of these species may not be directly related to "take" but may be loss or conversion of critical spawning habitat. Protection of spawning habitats requires public support and recognition of the threats that affect these critical areas and enforcement by natural resource agencies. Volunteers frequently are the most passionate and articulate protectors of wildlife species and habitats. A system of marine protected areas (MPAs) can provide comprehensive protection to ecosystems and species over a broad geographic area (Halpern et al., 2008), but additional efforts may be necessary for beaches that fall outside these boundaries and have multiple uses.

From the prescient Japanese who identified their kusafugu puffers as natural treasures more than 40 years ago, to the citizen scientists taking data as Grunion Greeters in California and the Capelin Observer Network in Canada, there is abundant evidence that public interest in beach-spawning fishes is widespread and sustainable. The impact that local observers can have on policies and management is immeasurable and irreplaceable. Data provided by citizen scientists and volunteers is vital to increase the knowledge of these understudied animals. However, the lack of funding for training, support, and recognition of volunteers makes it difficult to maintain consistent monitoring efforts over long periods of time (Devictor et al., 2010).

The local ecological knowledge of fishermen, naturalists, teachers, and other non-scientists is necessary to improve local and global management of coastal resources. The contributions of volunteer observers and resource agencies have improved the outlook for beach-spawning species at many locations across the globe. This book is dedicated to everyone who has shared the amazement and thrill of the natural spectacle of beach-spawning fishes. Everyone who has seen these fishes, or studied these fishes, or even tried unsuccessfully to see them, has a story they want to share with a smile. Here's hoping that these fishes continue to make their inspiring leaps of faith onto beaches, and that people continue to be awed and fascinated by this evocative behavior for many spawning seasons yet to come.

REFERENCES

Boulay, C. (2011). *Capelin Observers Network Observer Kit 2012*. Fisheries and Oceans Canada. http://www.qc.dfo-mpo.gc.ca/signaler-report/roc-con/2011-2012/Observer_Kit_2012.pdf.

California Marine Life Protection Act Initiative. (2009). *Regional Profile of the MLPA South Coast Study Region (Point Conception to the California–Mexico Border)*. Sacramento, CA: California Natural Resources Agency. http://www.dfg.ca.gov/marine/mpa/scpro-file.asp.

Department of Conservation, New Zealand. (2011). *The Whitebaiter's Guide to Whitebait*, 3 pp. http://www.doc.govt.nz/Documents/parks-and-recreation/activity-finder/fishing/whitebait-brochure-2011.pdf.

Devictor, V., Whittaker, R. J. & Beltrame, C. (2010). Beyond scarcity: citizen science programmes as useful tools for conservation biogeography. *Diversity and Distributions* 16, 354–362. DOI: 10.1111/j.1472-4642.2009.00615.x.

Dugan, J. E., Hubbard, D. M., McCrary, M. D. & Pierson, M. O. (2003). The response of macrofauna communities and shorebirds to macrophyte wrack subsidies on exposed sandy beaches of Southern California. *Estuarine, Coastal and Shelf Science* 58, 25–40.

Feder, H. M., Turner, C. H. & Limbaugh, C. (1974). *Observations on Fishes Associated with Kelp Beds in Southern California. Fish Bulletin 160*, State of California Resources Agency, Department of Fish & Game. http://content.cdlib.org/view?docId=kt9t1nb3s h&brand=oac4.

Frank, K. T., Leggett, W. C., Petrie, B. D., Fisher, J. A. D., Shackell, N. L. & Taggart, C. T. (2013). Irruptive prey dynamics following the groundfish collapse in the Northwest Atlantic: an illusion? *ICES Journal of Marine Science*. DOI: 10.1093/icesjms/fst111.

Frank, K. T., Petrie, B., Fisher, J. A. D. & Leggett, W. C. (2011). Transient dynamics of an altered large marine ecosystem. *Nature* 477, 86–89.

Haggerty, J. H. (2007). "I'm not a greenie but ....": environmentality, eco-populism, and governance in New Zealand: experiences from the Southland whitebait fishery. *Journal of Rural Studies* 23, 222–237.

Halpern, B. S., McLeod, K. L., Rosenberg, A. A. & Crowder, L. B. (2008). Managing for cumulative impacts in ecosystem-based management through ocean zoning. *Ocean & Coastal Management* 51, 203–211.

Hansen, M. (2013). Ukulele music, rubber duckies and a toadfish queen: Marin celebrates Independence Day. *Marin Independent Journal* 7.4.2013. http://www.marinij.com/ci_23601126/ukelele-music-rubber-duckies-and-toadfish-queen-marin.

Harzen, S. E., Brunnick, B. J. & Schaadt, M. (2011). *An Ocean of Inspiration: The John Olguin Story*. Toronto: Rocky Mountain Books.

Hikari City Board of Education, Culture, and Lifelong Learning Division. (2012). http://www.city.hikari.lg.jp/bunka/kusahugu25.html.

Hjermann, D. O., Ottersen, G. & Stenseth, N. C. (2004). Competition among fishermen and fish causes the collapse of the Barents Sea capelin. *Proceedings of the National Academy of Sciences* 101, 11679–11684. http://www.pnas.org/cgi/doi/10.1073/pnas.0402904101.

Hubbs, C. L. (1916). Notes on the marine fishes of California. *University of California Publications in Zoology* 16, 13, 153–169.

James, J. J. R. (2000). From beaches to beach environments: linking the ecology, human use, and management of beaches in Australia. *Ocean and Coastal Management* 43, 495–514.

Johnson, P. B., Martin, K. L., Vandergon, T. L., Honeycutt, R. L., Burton, R. S., & Fry, A. (2009). Microsatellite and mitochondrial genetic comparisons between northern and southern populations of California grunion *Leuresthes tenuis*. *Copeia* 2009, 467–476. DOI: 10.1643/CI-07-253.

Love, M. S. (2011). *Certainly More Than You Want to Know about the Fishes of the Pacific Coast: A Postmodern Experience*. Santa Barbara, CA: Really Big Press.

Marine Conservation Society, United Kingdom. (2014). Whitebait. http://www.fishonline.org/fish/285/whitebait (accessed May 2014).

Martin, K. L. M., Heib, K. A. & Roberts, D. A. (2013). A Southern California icon surfs north: local ecotype of California grunion, *Leuresthes tenuis* (Atherinopsidae) revealed by multiple approaches during temporary habitat expansion into San Francisco Bay. *Copeia* 2013, 729–739.

Martin, K. L. M., Moravek, C. L., Martin, A. D. & Martin, R. D. (2011). Community based monitoring improves management of essential fish habitat for beach spawning California grunion. *Bulletin de l'Institut Scientifique* 6, 65–72.

Martin, K., Speer-Blank, T., Pommerening, R., Flannery, J. & Carpenter, K. (2006). Does beach grooming harm grunion eggs? *Shore & Beach* 74, 1, 17–22.

Martin, K., Staines, A., Studer, M., Stivers, C., Moravek, C., Johnson, P. & Flannery, J. (2007). Grunion Greeters in California: beach spawning fish, coastal stewardship, beach management and ecotourism. In *Proceedings of the 5th International Coastal & Marine Tourism Congress: Balancing Marine Tourism, Development and Sustainability* (Lück, M., Gräupl, A., Auyong, J., Miller, M. L. & Orams, M. B., eds.), 73–86. Auckland: New Zealand Tourism Research Institute.

McDowall, R. M. (1996). *Managing the New Zealand Whitebait Fishery: A Critical Review of the Role and Performance of the Department of Conservation*, No. 32, NIWA Science and Technology Series. Wellington, NZ: National Institute of Water and Atmosphere Research Ltd.

McDowall, R. M. (2006). Crying wolf, crying foul, or crying shame: alien salmonids and a biodiversity crisis in the southern cool-temperate galaxioid fishes? *Reviews in Fish Biology and Fisheries* 16, 233–422. DOI: 10.1007/s11160-006-9017-7.

Middaugh, D. P. (1981). Reproductive ecology and spawning periodicity of the Atlantic silverside, *Menidia menidia* (Pisces: Atherinidae). *Copeia* 1981, 766–776.

Moulton, L. L. & Penttila, D. E. (2001). *Field Manual for Sampling Forage Fish Spawn in Intertidal Shore Regions*. San Juan County Forage Fish Assessment Project. Olympia: Washington Department of Fish and Wildlife.

Myers, R. A. & Worm, B. (2003). Rapid worldwide depletion of predatory fish communities. *Nature* 423, 280–283.

Nakashima, B. & Wheeler, J. P. (2002). Capelin (*Mallotus villosus*) spawning behavior in Newfoundland waters: the interaction between beach and demersal spawning. *ICES Journal of Marine Sciences* 59, 909–916. DOI: 10.1006/jmsc.2002.1261.

NBC Today Show. (1986). "Humming Toadfish Are the Buzz of Sausalito." 6/16/1986. New York, NY: NBC Universal. https://archives.nbclearn.com/portal/site/k-12/browse/?cuecard=40085.

Pacific Fishery Management Council (2014). Ecosystem Initiative 1: Protecting Unfished and Unmanaged Forage Fish Species. http://www.pcouncil.org/wp-content/uploads/I1a_ATT1_Eco_Initiative1_forage_APR2014BB.pdf (accessed May 2014).

Pauly, D., Christensen, V., Dalsgaard, J., Froese, R., & Torres, F. (1998). Fishing down marine food webs. *Science* 279, 860–863.

Penttila, D. E. (1995a). Investigations of the spawning habitat of the Pacific sand lance, *Ammodytes hexapterus*. In *Puget Sound Research'95 Conference Proceedings*, 855–859. Olympia, WA: Puget Sound Water Quality Authority.

Penttila, D. E. (1995b). The WDFW's intertidal baitfish spawning beach survey project in Puget Sound. In *Proceedings of the Puget Sound Research-95 Conference*, Vol. 1, 235–241. Puget Sound Water Quality Authority, Olympia, WA.

Penttila, D. E. (2007). *Marine forage fishes in Puget Sound*. No. TR-2007-03, 23 pp. Olympia, WA: Washington Department of Fish and Wildlife.

Pew Charitable Trusts. (2014). Pacific Fish Conservation Campaign. http://www.pewenvironment. org/campaigns/pacific-fish-conservation-campaign/id/85899360413/resources (accessed May 2014).

Pikitch, E. K. (2012). The risks of overfishing. *Science* 338, 474–475. DOI: 10.1126/science.1229965.

Pikitch, E. K. et al. (2014). The global contribution of forage fish to marine fisheries and ecosystems. *Fish and Fisheries* 15, 43–64. DOI: 10.1111/faf.12004.

Quinn, T., Krueger, K., Pierce, K., Penttila, D., Perry, K., Hicks, T. & Lowry, D. (2012). Patterns of surf smelt, *Hypomesus pretiosus*, intertidal spawning habitat use in Puget Sound, Washington State. *Estuaries and Coasts*. DOI: 10.1007/s12237-012-9511-1.

Robards, M. D., Anthony, J. A. & Piatt, J. F. (2002). Growth and abundance of Pacific sand lance, *Ammodytes hexapterus*, under differing oceanographic regimes. *Environmental Biology of Fishes* 64, 429–441.

Roberts, D., Lea, R. N. & Martin, K. L. M. (2007). First record of the occurrence of the California grunion, *Leuresthes tenuis*, in Tomales Bay, California: a northern extension of the species. *California Fish & Game* 93, 107–110.

Rose, G. A. (2005). Capelin (*Mallotus villosus*) distribution and climate: a sea "canary" for marine ecosystem change. *ICES Journal of Marine Science* 62, 1524–1530.

Smith, D. R. & Michels, S. F. (2006). Seeing the elephant: importance of spatial and temporal coverage in a large-scale volunteer-based program to monitor horseshoe crabs. *Fisheries* 31, 485–491.

Smith, L. (2012). Anchovies bounce back but whitebait are struggling, says guide. *Fish2fork News*, February 16, 2012. http://www.fish2fork.com/news-index/Anchovies-bounce-back-but-whitebait-are-struggling-says-guide.aspx.

Sutter, A. (2012). The humming toadfish story. *MarineScope Sausalito*, 7/11/2012. http://www.marinscope.com/sausalito_marin_scope/opinion/article_e009796f-01cc-5d1c-a60a-4e9bd75afe22.html.

Templeman, W. (1948). The life history of the capelin (*Mallotus villosus*) in Newfoundland waters. *Bulletin of the Newfoundland Government Laboratory at St. John's* 17, 1–155.

Thedinga, J. F., Johnson, S. W. & Mortensen, D. G. (2006). Habitat, age, and diet of a forage fish in southeastern Alaska: Pacific sandfish (*Trichodon trichodon*). *Fisheries Bulletin* 104, 631–637.

Uno, Y. (1955). Spawning habit and early development of a puffer, *Fugu* (*Torafugu*) *niphobles* (Jordan et Snyder). Journal of Tokyo University Fisheries 42, 169–183.

Vandepeer, F. & Methven, D. A. (2007). Do bigger fish arrive and spawn at the spawning grounds before smaller fish: Cod (*Gadus morhua*) predation on beach spawning capelin (*Mallotus villosus*) from coastal Newfoundland. *Estuarine, Coastal, and Shelf Science* 71, 391–400.

Walker, B. W. (1949). The periodicity of spawning by the grunion, *Leuresthes tenuis*, an atherine fish. Dissertation, University of California, Los Angeles.

Walker, B. W. (1952). A guide to the grunion. *California Fish & Game* 38, 409–420.

Index

A

Adinia xenica, 44
Airbreathing fishes. *See* Amphibious fishes
Allosmerus elongatus, 10
Alternative reproductive tactics (ART), 36.
    See also Mating behaviors and
    systems
Alticus kirki, 39–40
Ammodytes hexapterus, 64, 127, 181
Amoring, shoreline, 158
Amphibious fishes. *See also* Beach-spawning
    fishes; *specific fishes*
  active emerger fishes, 83
  adaptations, 80–81, 86, 87
  adaptive evolutionary changes, 86, 87
  adult fishes *versus* eggs and embryos, 94
  aerial gas exchanges, 89–90
  carbon dioxide exchanges, 81
  evolution, 20
  gas exchange, 20, 25
  habitats, 79, 87
  hypoxic conditions, 21, 22, 24, 25–36, 79–80,
    83, 90, 91, 96, 145–146
  metabolic rate, 87–88
  mudskippers; *see* Mudskippers
  overview, 79
  oxygen exchanges, 81
  passive remainer fishes, 81, 82, 89
  physiological consequences of, 94
  respiratory acidosis, 85
  respiratory structures, 84–85, 85–86, 87
  salinity of water; *see* Salinity, water
  skippers; *see* Mudskippers; Rockskippers
  species variety, 20
  substrates, 79
  temperature of water; *see* Water temperature
Anableps anableps (Anablepidae), 47
Anadromy, 8, 10, 68–69, 114
Andamia tetradactyla, 40, 123
Anoxia-induced quiescence, 146–147
Antarctica, 26
Apletodon dentatus, 124
Aquatic hypoxia, 143–144
Artedius species, 38
Ascelichthys rhodorus, 38
Atheriniformes, 8
Atherinops affinis, 10
Atherinopsidae, 8, 9
Atlantic silversides. *See under* Silversides

Austrofundulus limnaeus, 124
Axoclinus nigricaudus, 36, 40–41

B

Batrachoididae, 6
Beach grooming, 163
Beach replenishment, 164, 166
Beach-spawning fishes. *See also* Freshwater
    fishes; Subtidal fishes; *specific fishes*
  1, 4–5
  advantages of beach-spawning, 34
  air breathing; *see* Amphibious fishes
  amphibious fishes; *see* Amphibious fishes
  antecedent behavior, 11
  aquatic spawning, 34
  behavior, 4, 5–6, 11, 69
  cannibalization of eggs; *see* Cannibalism
  conception; *see* Reproductive cycle, beach-
    spawning fish
  death of adults, post-spawning, 6–7
  demersal eggs; *see* Demersal eggs
  depth of spawn, 6, 7
  dessication, impact of, 4
  egg placement, tidal heights, 125
  eggs; *see* Reproductive cycle, beach-
    spawning fish
  embryo incubation; *see* Reproductive cycle,
    beach-spawning fish
  embryos; *see* Reproductive cycle, beach-
    spawning fish
  estuaries; *see* Estuaries, spawning
  evolution of, 4–5, 6, 9
  gas bladders of, 93–94
  gravel beaches, 26, 60–65
  habitat diversity, 1, 4, 8, 11, 24
  habitat, changes to, 167–168
  hatchlings; *see* Hatchlings
  heat stress, 4
  intertidal species, 4–5, 18, 19–20, 37–43,
    57–58, 60, 67–68, 80–84, 89, 113, 125;
    see also specific fishes
  length of spawning behavior, 6
  lineages, independent origins of behavior
    of, 5–6
  mating behaviors and systems; *see* Mating
    behaviors and systems
  migration; *see* Migration
  nesting; *see* Nesting, beach-spawning fishes
  overview, 1, 4

9 780367 659059